ANIMAL BEHAVIOUR · VOLUME 1

CAUSES AND EFFECTS

ANIMAL BEHAVIOUR

A SERIES EDITED BY

T.R. HALLIDAY
Department of Biology
The Open University

AND

P.J.B. SLATER
School of Biology
University of Sussex

ANIMAL BEHAVIOUR · VOLUME 1

CAUSES AND EFFECTS

EDITED BY T.R. HALLIDAY
AND P.J.B. SLATER

W. H. Freeman and Company
New York / San Francisco

© 1983 by Blackwell Scientific
Publications Ltd

Published in the United States in 1983
by W.H. Freeman and Company,
41 Madison Avenue, New York,
NY 10010. First published in Great
Britain in 1983 by Blackwell Scientific
Publications Ltd

Printed and bound in Great Britain

Library of Congress
Cataloging in Publication data

Main entry under title:

Animal behaviour.

Includes bibliographies and indexes.
Contents: v. 1. Causes and effects—
v. 2. Communication—v. 3. Genes,
development and learning.
1. Animal behavior. I Halliday,
Tim, 1945–. II. Slater, Peter J.B.
[DNLM: 1. Behavior, Animal—
Physiology. 2. Motor activity—
Physiology. 3. Motivation.
W1 AN228DZ v. 1/QL 751 C374]
QL751.A649 1983   591.51   83–5497
ISBN 0–7167–1578–3 (v. 1)
ISBN 0–7167–1579–1 (v. 1: pbk.)

CONTENTS

v

SERIES INTRODUCTION

As Niko Tinbergen, one of the founders of ethology, pointed out, if one asks why an animal behaves in a particular way, one could be seeking any one of four different kinds of answer. One could be asking about the evolutionary history of the behaviour: why did it evolve to be like it is? One could be asking about its current functions: through which of its consequences does natural selection act to keep it as it is? Thirdly, one might be interested in the stimuli and mechanisms that lead to the behaviour being performed: what causes it? Finally, one might be asking about development: how does the behaviour come to be as it is during the life of the individual animal? A complete understanding of behaviour involves investigation of all these questions, but in recent years there has been a tendency for ethologists to specialise in one or other of them. In particular, the functional analysis of behaviour has almost become a separate discipline, variously called behavioural ecology or sociobiology. This fragmentation of the subject is unfortunate, because all its facets are important and an integrated approach to them has much to offer.

Our approach in these books has been a more wide-ranging one than has been common in recent texts, with attention to all the kinds of explanation that have traditionally been the concern of ethologists. Aimed at students, each volume will provide a comprehensive and up-to-date review of a specific area of the subject in which there have been important and exciting recent developments. It is no longer easy for a single author to cover the whole field of animal behaviour with full justice to all its aspects. By asking specialists to write the chapters, we have tried to overcome this problem and ensure that recent developments in each area are fully and authoritatively covered. As editors, we have endeavoured to make sure that there is continuity between the chapters and that no significant gaps have been left in the coverage of the theme specific to each book. We hope that students who are inspired to further study by what they read will find the Selected Reading recommended at the end of each chapter a useful guide, as well as the more specific references which are gathered together at the end of each book.

We thank Bob Campbell and Simon Rallison of Blackwell Scientific Publications for their help and encouragement throughout the preparation of these books, Clare Little of Oxford Illustrators for her fine work on the illustrations and, most important of all, our authors for their readiness to accept a well-defined brief, to meet deadlines, and to accept our editorial changes and promptings.

T.R.H.

1983   P.J.B.S.

ACKNOWLEDGMENTS

Both of the editors and, in most cases, some of the authors of other chapters have commented on each chapter in draft. In addition, thanks are due to Alastair Houston and Tim Roper for their comments on an early draft of Chapter 5.

INTRODUCTION

What causes an animal to behave in a particular way? As Aristotle realised, this question can be posed at several different levels. Those that we might identify at the present day would include the evolutionary or functional question of why a behaviour pattern is included in the animal's repertoire, and also the developmental question of what influences during the lifetime of the individual have led it to behave in this particular manner. This second approach is dealt with in Volume 3 of this series. In this volume, however, we are concerned with a third class of cause, which Aristotle termed the efficient cause and which we now often refer to as the immediate cause of an animal's actions. This is the most common usage of the term cause, and covers only factors which lead to behaviour in the short term, excluding more long-term antecedents such as selective forces, inherited tendencies and aspects of past experience. Here, then, we are asking what external stimuli, internal states and mechanisms lead to the performance of the behaviour that we see.

Those who study the immediate causes of behaviour can take many different approaches, as the chapters of this book will make clear. Perhaps the most fundamental distinction is between those who remain at the behavioural level and those who seek answers in physiology. To study behaviour at its own level is to treat the animal as a 'black box', modifying the stimuli that impinge upon it and seeing how these affect its behavioural output without being concerned too deeply with the exact nature of the intervening mechanisms. Both external stimuli, such as length of day or presence of other animals, and internal stimuli, such as hormone levels or gut content, can be altered in this way and just how they affect behaviour may be studied. This enables behaviour to be predicted, to be changed and its causes to be understood without any direct knowledge of the mechanisms within the animal for translating the various inputs into the output. To others, answers such as this are unsatisfying: they want to know exactly what is going on to produce the behaviour in terms of hardware, to

1

analyse the neural circuitry between sense organs and muscles responsible for it. Only then will they feel that the behaviour has been 'understood'. This is the essence of the reductionist approach: seeking to explain each level of organisation in terms of that below it. At an extreme it could be claimed that human psychology should ultimately be understandable in terms of the behaviour of atoms. This is obviously absurd, but is it absurd simply because it is impracticable?

Reductionism does have its enthusiastic adherents, and there are those who argue that studying the causation of behaviour is ultimately a matter for neurophysiologists (e.g. Wilson 1975). However, there are enormous pitfalls. Quite apart from the sheer complexity of the matter, is the point that each level of organisation tends to have so-called 'emergent properties' which could not be predicted from the level beneath. As Marr (1982) put it, in a stimulating review of levels of approach to understanding vision:

> 'Almost never can a complex system of any kind be understood as a simple extrapolation from the properties of its elementary components'.

In contrast to this we can take the following remark by Barlow (1972), made at a time when our knowledge of the nervous system was advancing rapidly and appeared to hold the key to most things:

> 'A description of that activity of a single nerve cell which is transmitted to and influences other nerve cells and of a nerve cell's response to such influences from other cells, is a complete enough description for functional understanding of the nervous system'.

Barlow's remark was clearly intended to provoke research and discussion but it embodied the optimism of that time, when sensory physiologists were going deeper and deeper into the nervous system and it seemed that they might almost emerge at the other side with an 'explanation' of behaviour. Such an enterprise is, however, doomed to failure for the practical reason that translating between levels is a very complex process and for the more theoretical reason that properties emerge so that questions of interest at one level may have no counterpart at that beneath. This is not to say that the neural mechanisms underlying behaviour are of no interest, it is just to point out that they may be extremely hard to unravel (as is proving to be the case) and that understanding them is only part of the answer to understanding the causes of behaviour.

A point related to the merits of reductionism is the question of how mechanistic behaviour is. The answer to this question from someone who studies behaviour depends strongly on the complexity of the animals they work on. Certain aspects of behaviour, like reflexes in ourselves and in other animals, are virtually automatic. The neural machinery is simple, perhaps involving only two or three cells, and the response is near constant and highly predictable. In some animals, especially amongst invertebrates with small repertoires of rather fixed behaviour patterns, much of the behaviour may be like this. For example, it may not be too great a simplification to summarise the behaviour of a spider confronted by a visual stimulus by the rule: if it is small eat it, if it is large run away from it and if it is intermediate in size mate with it. However, as animals become more complex, with numerous possible modes of behaviour, these cannot be subject to simple rules which do not change with time. In other words, the problem of changing motivation crops up. To those who wish to study behaviour from a mechanistic point of view this can be a nuisance and more consistent animals must be used. Take Ewert (1974), for example, justifying his choice of the toad for experiments on visually guided behaviour:

'In response to specific stimuli one can repeatedly elicit predictable reactions, such as snapping at prey, fleeing from an enemy, clasping during courtship and making particular wiping motions after tactile stimulation. (The fickle European frog, in contrast, undergoes short-term changes in motivation and is not suitable for behavioural experiments.)'

Where one person's interest ceases, another's begins; to those interested in understanding the intricacies of motivation, the frog has virtues that the toad lacks! Indeed, for most vertebrate behaviour there are short-term changes in responsiveness which make it hard to obtain consistent results. Some such changes may be for relatively trivial reasons, such as adaptation of sense organs to the relevant stimuli or fatigue in appropriate muscles, but often these can be excluded as the senses and muscles are used for other activities. Some more central change must then be responsible. It may not be surprising that an animal which has just eaten to satiation is not interested in food, but why should a male stickleback chase a rival one day but pay scant attention to him the next? Understanding such changes is the bread and butter of motivation study and, at least amongst vertebrates, changes in motivational state are a very important determinant of whether or not an animal will respond when a stimulus is presented to it.

This book is concerned with mechanisms and with motivation, but it is not a book about physiology. Reference will, of course, be made to physiological factors, such as levels of hormones, glucose and ions in the blood, and to characteristics of the nervous system. But the focus throughout is on behaviour, and these subjects are raised only where they help us to understand behaviour rather than as central topics. While our authors offer very different views and perspectives on many issues, they are united in approaching behaviour at its own level rather than simply as a projection of physiology.

The first three chapters are concerned with the more physical aspects of behaviour, Land with sensory systems, Dawkins with movement patterns and Collett with how the two are coordinated so that the animal behaves in an appropriate fashion. Animals can glean information about the world in which they live in numerous different ways, for various sorts of stimuli impinge upon them and can be used to guide their actions. Most commonly, pressure waves stimulate hair cells within the ears to create the sensation of sound, electromagnetic radiation over a limited range of frequencies is picked up by eyes and used to create a visual picture of the world, and chemicals, especially from other animals and from food, are sensed by smell and by taste. The main aim of Chapter 1 is to survey the properties of the senses and show how animals can use them to extract information about their surroundings. In common with subsequent chapters, Land takes a comparative approach. When discussing the senses this is especially interesting for two reasons. First is the remarkable variety of ways that evolution has generated for picking up information of a particular sort and translating it into a form usable by the nervous system. Eyes make a good example: though all perform similar tasks, for subtle reasons related to their exact uses and the medium in which they operate as well as the phylogenetic relationships of their possessors, they come in many different designs. Second is the fact that animals are by no means limited in their senses to those we possess ourselves. They may sense a broader range of frequencies that we can, as with bees that see the ultraviolet patterns on flowers, snakes that sense the infrared radiation from their warm-blooded prey and bats that orientate using high-frequency sounds far beyond the limit of our hearing. In other cases the sense may be of a different quality altogether, as with the magnetic sense of birds, bees and bacteria and the electrical sense of some fish.

Having gleaned from their senses what is going on in the world around them, to what use do animals put the information? This is the main theme of Collett's chapter. It may seem surprising that he starts with examples from single cells, as many people would not think of such lowly organisms as 'behaving'. It is, however, salutary to consider the sophisticated and adaptive ways in which bacteria and protozoa can alter their movements in relation to stimuli from the world in which they live, without so much as a nerve or muscle to assist them! Nevertheless, one can only get so far by simple reactions to whether or not a particular sense is stimulated. In most animals subtle changes in the sense organs lead to equally subtle changes in behaviour, detection of the appropriate sensory events involving collation of information from many receptors and the response involving particular combinations of many muscles which may, at other times, take part in quite different actions. Collett discusses these relationships. In some cases they appear to involve the straightforward application of simple rules which translate sensory cues into appropriate action so that behaviour can be matched to environment elegantly and economically. Only with the intervention of prying scientists do things go awry and so give us an insight into the rules that guide the animal's actions. In other cases, especially amongst mammals, one must invoke more than just simple rules: here several senses may fall into register with one another within the brain to give the animal a multimodal picture of where it stands in relation to its world, and the animal must learn about that world and retain a mental map of it to match its actions to it. Rudimentary stimulus–response reactions will not guide an animal to an unseen water-hole or to the best place of refuge from a predator regardless of where it happens to be on its territory when the danger arises.

In considering the motor activities of animals, Dawkins moves us into a realm which has for long been central to ethology, and provides a bridge to the chapters on motivation which follow. If we watch an animal moving around in its environment and going about its daily business of feeding, drinking and grooming, it is easy to be overawed by the complexity of it all. How can one start to analyse something so diverse? One of the great contributions of the early ethologists was to recognise that some order existed in the midst of this diversity, and that animals had a repertoire of stereotyped behaviour patterns that could be spotted each time they occurred and were the same throughout a species. Thus all horses walk by moving the two legs which are diagonally opposite each other at the same time, but all giraffes, surprisingly for an animal so tall, move the two legs on the same side at once, swaying from side to side to keep their centre of gravity above the pair on the ground. Later work has revealed that various attributes of such Fixed Action Patterns, including their fixity, were not as hard and fast as had been supposed. Dawkins discusses these problems and others concerned with defining units of behaviour. She then goes on to look at the organisation of these basic units into higher-order sequences of actions and at theories underlying this organisation, notably the idea that some behaviour is goal-directed and the idea that related actions fit together in an organised manner because there is a hierarchy of control. The idea of hierarchical organisation was espoused most strongly by Tinbergen (1951) who, in a first flush of enthusiasm for the possibility of making links between behaviour and the nervous system, devised a hierarchical model incorporating centres which he hoped might have physiological manifestation within the brain. As Dawkins shows, life is not that simple and, while advances have certainly been made in understanding the neurophysiological basis of behaviour patterns and the sequences into which they fall, she argues strongly for the 'whole animal' approach and for the understanding of behaviour this can give us.

While single actions may appear in much the same form every time an animal performs them, sequences of different activities are seldom repeated in exactly the same order. In other words, behaviour is probabilistic rather than deterministic in its structure: a good reason why it appears to be so complex. This is also a good reason why changing motivation often has to be invoked as a factor to explain the structure of behaviour.

Although the study of motivation deals with many different topics, it is most convenient to split them up into the factors affecting the occurrence of single systems of behaviour, as discussed by Halliday in Chapter 4, and the interactions between systems, which McCleery covers in Chapter 5. Both these subjects have received a great deal of attention from ethologists, and views on them have changed markedly over the past few decades. Once again, rather sweeping generalisations by early ethologists provided a stimulus for research which, in its turn, showed just how inadequate these theories were. As far as the motivation of sets of related behaviour patterns was concerned, the first theory proposed was Lorenz's (1950) 'psychohydraulic' model known more colloquially as Lorenz's water-closet. This pictured 'action specific energy' as accumulating and being expended like water in a tank, and is discussed as a starting point by both Halliday and Toates. It is a naive first approximation to what may perhaps be going on in some systems, but it trips up on a number of important facts. Two of these deserve mention here. First is the assumption that different behavioural systems are likely to have similar underlying mechanisms. We know that, like Lorenz's model, the tendency to feed and drink rises with time elapsed since these patterns were last performed. We would expect this, as these actions correct physiological deficits which are bound to accumulate. But would we expect the same of grooming, sex, aggression or egg-rolling in geese? In his book *On Aggression* (1966) Lorenz clearly does have such a model in mind. However, recent research has shown the application of such general theories to be dangerous: one of the wonderful things about animal behaviour is how finely tuned it has become through natural selection to the requirements of each species. Different systems may behave quite differently depending on the exact requirements of their possessors. A second problem with Lorenz's model is that it makes no provision for the behaviour to be influenced by feedback from its own consequences. As Toates points out in Chapter 6, the application of control theory to behaviour patterns such as feeding and drinking has been especially fruitful, and feedback diagrams help us to predict behaviour with a fair degree of accuracy. Except in the most stereotyped of behaviour patterns, animals continuously monitor and modify their behaviour in the light of its consequences: they stop drinking, for example, when they have accumulated sufficient rather than, as Lorenz's model would predict, when they have taken a certain number of gulps.

Generally speaking, human beings can walk and chew gum at

the same time. Amongst most animals, however, it is unusual to find the performance of two different behaviour patterns proceeding simultaneously. A bird tends to be either flying or preening or feeding or singing, although there are of course exceptions. This makes the analysis of behaviour much easier but it leads to a whole range of problems in understanding motivation. How should an animal decide what to do at a particular time? If it is hungry and itchy, there is a prospective mate nearby and a predator is looming, should it escape, mate, feed or groom? These are the sorts of problems which McCleery discusses in Chapter 5. For some species a fixed hierarchy of priorities provides a reasonable description of changing motivation: in the snail *Pleurobranchaea*, for example, egg laying suppresses feeding, and escape takes priority over both. In most species, however, priorities change and different acts are more likely at different times. Earlier ethologists pictured drive systems vying with each other for expression, with the possibility that where two were in conflict a third might be disinhibited, so that unexpected behaviour appeared in the form of a displacement activity. But internal drives and conflicts between them are not very fruitful hypotheses because they can account for almost anything one might observe: if you expect the animal to groom it is doing so because its grooming drive is high, if not it is doing so because other drives are in conflict and grooming has been disinhibited. Because of such difficulties with intervening variables which cannot be observed, drive theories tend to have fallen from favour. However, the problems remain and McCleery describes the more sophisticated and precise modern theories which have been formulated to cope with them.

The problems of motivation, whether they deal with the occurrence of a single behaviour pattern such as feeding or with sequences of different actions, are complex, some would say intractable. Devising models of how systems may work has always been helpful, regardless of whether or not the model is thought of as having any physiological reality. Building models can be quite an eye-opener and is now an easy matter for anyone with a little knowledge of computer programming. We know the system has certain properties and we build these into our model. How does it behave? Usually not very well at first, but we can modify it in one way or another to make it more realistic. We can incorporate further features if these prove necessary but sometimes, to our surprise, we need not do so: unexpectedly a property of behaviour emerges from the model without having been predicted, or the model behaves in a particular way when tested and we find that the same manipulation of a real animal has the same effect. Models are a marvellous way of helping us to think about really complex things and, by moving back and forth from animal to computer, we can discover just what is needed to predict behaviour when this might otherwise be beyond our capabilities. Toates shows some of the potential of the modelling approach, referring back to some of the problems discussed by Halliday and McCleery. His discussion of drinking will be a revelation for anyone who thought that all one need postulate is thirst!

CHAPTER 1
SENSORY STIMULI AND
BEHAVIOUR

M.F. LAND

1.1   Introduction

It is the information supplied by senses that guides an animal's present and future actions, and it is the variety, quantity and quality of this information that to a large extent determine the complexity and flexibility of the animal's behavioural repertoire. In the simplest cases a sense organ may supply no more than a statement about the intensity of a particular form of energy. The paired eyespots of a planarian, for example, tell the animal which direction is the lighter, but no more, and the action resulting from this knowlege is a movement either towards or away from the lighter region. As the next chapter will show, the simplest of information can thus be used to guide an animal to a suitable habitat. Similarly, a sudden decrease in light intensity may be used by a fanworm or a clam to indicate the imminent presence of a predator, and the animal will withdraw into its tube, or close its valves. Although there are many examples of behaviour of this kind, where it seems possible to establish a straightfoward one-to-one link between a change in a stimulus and a recognisable kind of motor activity, with more sophisticated sense organs and phylogenetically more advanced animals this is usually not possible. There are several reasons for this, and they are all pertinent to the problem of how animals use their senses.

In the first place, most sense organs supply far more information than can be made use of at any one time (consider, for example, that the human eye alone has about 120 million receptors, each capable of allowing discrimination between at least ten intensity steps). At any one time, therefore, most incoming sensory information is redundant or unusable. There can be no question of a simple reflex link between complex receptors and effectors, if only because an animal's motor system can only engage in one course of action at a time. Incoming messages must be rigorously selected, and this selection can take two forms. At one extreme, a particular stimulus configuration may have such importance, or 'salience', that its presence overrides everything else. Few animals with good eyesight fail to take some kind of evasive action when presented with a large looming object, one whose projection on the retina expands rapidly. Alarm calls and alarm pheromones similarly elicit urgent defensive action. On the other hand, most sensory events are not of this kind, and whether or not they come to affect motor action depends on the kind of activity in which the animal is engaged. A red breast or a bunch of red feathers generally produces a ferocious attack from the European robin (*Erithacus rubecula*) but not if the feathers are on its mate (Lack 1953); a bee learns the colour of a flower to which it will later return, but only during the three seconds preceding feeding, not during and not after feeding (Lindauer 1969); and a locust will stabilise its head with reference to an artificial horizon, but only when it is actively flying (Camhi 1974). These examples all show that stimuli which are effective on one occasion are ineffective on others, for reasons that may be related to hormonal state, or to the nature of the current or preceding activity. Information may also be assimilated, and used much later. Chaffinches (*Fringilla coelebs*), for example, may learn their song in the first year but do not sing it until the next (Thorpe 1961), and indigo buntings (*Passerina cyanea*) learn to pick out the pole star well before they need stellar navigation to migrate by (Emlen 1975). In humans it is likely that much of what is taken in is never used.

The second main reason why a straightforward reflex account of an animal's 'input–output' relations is inadequate is that it fails to take into consideration the fact that every action produces a new set of sensory consequences. A few steps, or even a small eye movement, radically change the retinal image. Thus not all sensory stimuli derive from events in the external environment itself. Formally, an animal is in a relationship with the environment which resembles a closed feedback loop, a point noted by von Uexküll in the 1920s (Fig. 1.1) and stated in more explicit terms in the 'reafference principle' of von Holst and Mittelstaedt (von Holst 1954). They make a distinction between 'exafferent' information related to actual events in the outside world—the kind of

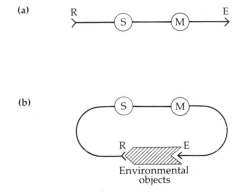

Fig. 1.1. von Uexkull's distinction between a simple reflex (a), and the 'closed loop' relationship (b) of an animal and its environment. R = receptor, E = effector, S = sensory neurons and M = motoneurons of the CNS. (After von Uexküll 1957.)

events that we usually refer to as 'stimuli'—and 'reafferent' information resulting from the animal's activities. One might think reafferent information would be especially uninformative and redundant, since it is largely predictable, but this is not always so. An animal trying to keep on a straight course, for example, makes use of the fact that rotational slip of the image across the eye indicates an involuntary turn, and it makes an immediate counter-rotation. This is the basis of the well-known optomotor response, in which an animal in a striped drum will turn its eyes, head or body in the same direction as the stripes, and it can be regarded as a visual feedback loop that stabilises both vision and locomotion. Interestingly, some electric fish show a similar phenomenon, moving when objects in their surroundings move and so change the pattern of the electric field around them (Heiligenberg 1977; see Fig. 1.9). This does not mean that all sensory input is, in any useful sense, feedback information, and for the escape responses of the worm or clam a push-button reflex description remains more apt. It does, however, mean that, for all its apparent precision, one has to be rather careful of the word 'stimulus'. It is not always easy to distinguish sensory events imposed on an animal from those of its own making.

A third, related, point is that reception is not usually a passive process but an active investigatory one. Bats explore their environment by emitting ultrasonic clicks into it and listening to the echoes, and if they detect an insect these clicks are produced more frequently (Sales & Pye 1974). Electric fish explore their world in a similar manner but by emitting electric pulses from the tail and monitoring on the head the distortions of the resulting electric field caused by objects in the surroundings (Lissmann 1963). An electric fish wishing to investigate an object more closely may bend its tail around it. In these two examples the inquisitive nature of the sense is obvious, because the animal must actually emit the physical energy required. However, even in senses where this is not the case, the sense organs are usually more than just passive receivers. Birds very obviously search their environment visually using head movements, and primates do the same with eye movements. Rabbits, like some bats, scan with their ears. Moths and many other insects search through odour plumes to locate the direction of the concentration gradient. It is often more appropriate to say that an animal seeks out stimuli than to say that it just receives them.

1.1.1   Classifying the senses

Since the time of Sherrington, early in this century, it has been usual to divide the senses into three kinds: enteroceptive, proprioceptive and exteroceptive (Schmidt 1978). Enteroceptors monitor the internal state of the animal—levels of oxygen, ions, temperature, blood sugar and so on. Proprioceptors either measure the relations of parts of the body to each other (muscle spindles, for example, monitor muscle length and hence limb extension), or they may measure self-generated movements of the animal as a whole (the semicircular canals of the vertebrate inner ear would be included in this category). The exteroceptors are then the sensory systems that monitor events outside the animal such as those received in humans by the classic five senses of vision, hearing, touch, taste and smell. There are many others in ourselves, skin temperature, vibration and pain for example, and others exist in other animals, two of the most remarkable being the ultrasonic sense of bats and electric senses of some fish. The term 'modality' is usually used interchangeably with 'sense', to mean the reception of a particular type of distinct physical or chemical stimulus. This chapter is concerned with the exteroceptive modalities, and the only caution that should be given before beginning a survey is to point out that exteroceptive and pro-prioceptive senses often overlap. One can use vision pro-prioceptively, for example to monitor the position of one's hand or steady oneself in space, and conversely muscle spindles and joint receptors can be used exteroceptively to judge the weight of a stone. As implied in the reafference principle, there is no very clear boundary between external and self-generated sensory events.

1.2 Sensory modalities

1.2.1 Vision

Of all the modalities, vision provides the most straightforward and accurate source of spatial information. Because light travels through air or water in straight lines and can be focused by the lenses of eyes, there exists on an animal's retina a pattern of light and dark that directly represents the two-dimensional projection of the corresponding pattern in the surroundings. The image is an unprocessed picture of the spatial distribution of reflecting objects in the world outside. The most elementary misconception about eyes, which stems from the way we ourselves see, is that the image on the retina is what animals actually perceive, and that the patterns of elements in it—trees, flowers, etc.—are available to the animal just because there is an image. The more we learn from physiology, and from computer-aided attempts to recognise patterns, the more complicated the process of extracting patterns seems to become. The image itself is as informative as a photograph in a closed album: it is the immense interpretative power of the brain behind the eye that makes possible, for us, the convincing translation of that image back into a representation of the world 'out there'. We cannot take for granted that the same facilities for interpretation exist in animals other than ourselves. These facilities must, indeed, be discovered by experiment. To give a single example, the lateral eyes of jumping spiders are based on a similar design to our own, with a good image projected on to a retina of about 10^4 receptors. These eyes are used only to see the movement of small objects, to which the animal turns, and they make no distinction between a fly, spider, human, car or bird; the turn is always the same. In contrast, the forward-pointing principal eyes do make distinctions, at least between prey, conspecifics and larger 'predatory' objects (Land 1972). Thus even in one animal there may be differences in interpretative ability between different pairs of eyes and the neural machinery they supply.

Eyes and images

Despite this caution, it remains the function of an eye to provide a usable image with which the nervous system can work. Images are produced in many different ways (Fig. 1.2), and convergent evolution has often occurred (Land 1980a). Thus one finds eyes built on the 'camera' plan, with a single lens and extended retina, in

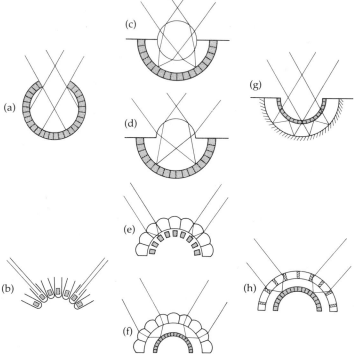

Fig. 1.2. Eight different ways of forming a resolved image. (a) Pigmented cup (planarians); (b) compound eye with receptors in simple pigment tubes (sabellid tube worms); (c) aquatic lens eye (fish, cephalopods); (d) corneal refracting surface (spiders, land vertebrates); (e) apposition compound eye (diurnal insects and crustacea); (f) superposition compound eye (nocturnal insects, krill); (g) simple mirror eye (scallops); (h) superposition eye using mirrors (shrimps, lobsters). Receptor cells are shown stippled. (From Land 1980a.)

cubomedusae, alciopid worms, strombid and heteropod snails, cephalopod molluscs and some insect larvae and spiders, as well as in all vertebrate groups. Scallops use a concave mirror rather than a lens to form an image, a mechanism that is almost unique. Compound eyes are principally found in arthropods, but they have also evolved independently in some bivalves (e.g. *Arca*) and a few tube-dwelling annelids (e.g. *Branchiomma*). Despite their superficial similarity, compound eyes are of several distinct types. Most diurnal insects have apposition eyes, in which each cluster of eight or nine receptors views the world through its own private lens, and the receptors in each cluster all look at the same solid angle of space. In superposition eyes, typical of nocturnal insects and deep-water crustaceans, the whole optical array produces a single erect image on a deep-lying retina, in a way much more like a camera eye than an apposition eye. The optics of superposition eyes are complicated and may involve an array of lenses (moths, krill) or mirrors (prawns, crayfish) (Land 1980b).

Resolution and sensitivity

The type of optical system employed makes little difference to the image as seen by the receptors, except for the minor technicality that it is inverted in camera eyes and erect in compound eyes. The two features that do affect performance whatever the type of eye are resolution and sensitivity. An eye's resolving power depends on the angle in space that individual receptors subtend, and this is typically around 1° in insects, and 1′ in the camera eyes of vertebrates and cephalopod molluscs. Compound eyes are at a disadvantage because of the small size of the individual lenses: diffraction theory states that the minimum resolvable angle of an optical system is *c.* λ/d radians, where λ is the wavelength of light (about 0.5 μm) and d the lens diameter. A 25 μm facet can thus only resolve 1/50 radian, or about 1°. A relatively coarse mosaic would be a disadvantage for very fine pattern discrimination, but not necessarily for guiding rapid flight where fine detail is not helpful. The resolution an eye provides can only be exploited if there is enough light. Light energy exists as indivisible photons, which are either absorbed entirely or not at all. In dim light, individual receptors may only be receiving a few photons per second—at the absolute threshold for human vision they may receive as few as one photon per receptor per 45 minutes! (Wald *et al.* 1962).

Since photon captures by receptors are randomly distributed in time, seeing in dim light is a statistical matter (Pirenne 1967; Land 1980a). One can calculate, for example, that for two receptors to signal a difference of 10% in the light they receive with a 95% chance of giving the right answer requires an average of about 800 photons per receptor. For a 1% difference this would be 80000. A white card at sunset will provide single receptors of a bee or a man with about the same numbers of photons, roughly 2500 per second, enough to discriminate a 10% but not a 1% difference in brightness. The point of this example is that despite the exquisite sensitivity of single receptors, which do indeed respond to single photons, the performance of all eyes is limited by photon fluctuations, and obviously this situation is worse the less light there is available. The eyes of nocturnal and deep-sea animals have evolved to project as many photons as possible on to the receptors by using either large lenses or superposition optics, and large receptors. Whether an eye has evolved for high spatial resolution, with many small receptors, or high sensitivity, with fewer larger receptors, it is bound to be large in order to accommodate these. It is perhaps not surprising that the largest known eye (37 cm diameter) was found in a deep-sea squid with a need for both high sensitivity and good resolution.

Colour and polarisation

The sensory potentialities of light extend beyond intensity and direction, the basic elements in a black-and-white image. Light has two other qualities: its wavelength distribution and the way in which it is polarised. In humans the visible spectrum in daylight extends from 400 nm (subjectively violet) to 800 nm (deep red), a minute part of the huge electromagnetic spectrum. This range is extended slightly in other animals. Most insects can see into the ultraviolet, down to 320 nm in bees (Menzel 1979), and this sensitivity is matched both by the ultraviolet reflecting patterns of many flowers, not visible to us, and also by ultraviolet patterns on some butterfly wings which serve as mate attractants (Eisner *et al.* 1969; see Fig. 1.3). At the other end of the visible spectrum, some snakes are sensitive to the infrared radiation emitted by the bodies of

Fig. 1.3. Ultraviolet patterns in flowers and insects. Normal light above, ultraviolet below. (From Eisner *et al.* 1969.)

warm-blooded animals, and they use this sense in prey capture. Interestingly, the eyes are not used for this, but a separate pair of pits that resemble poor pinhole cameras, situated between the eyes and nostrils (Gamow & Harris 1973). Infrared vision cannot be used by warm-blooded animals because the body itself is too strong an emitter. Colour vision in vertebrates and in insects is generally trichromatic; that is, there exist in the respective retinae three classes of receptor (called cones in vertebrates), each of which is maximally sensitive to a relatively narrow band of wavelengths. In humans these classes of cone have maximal sensitivities at 420 (blue), 534 (green) and 564 nm (yellow); in bees they are at 340 (U.V.) 450 (blue) and 540 nm (green). We can, of course, distinguish many shades of colour between these sensitivity maxima, as can bees, and this must be done by the brain comparing the relative extents to which each receptor class is stimulated. No one receptor can distinguish between intensity and wavelength on its own. At night, when the cones of most vertebrates are no longer active, the rods, with a maximal sensitivity in the blue-green (498 nm in humans) take over, and colour vision is lost. The capacity for colour vision seems to be more widespread than was formerly thought. It is present in almost all vertebrates that are neither strictly nocturnal nor inhabitants of the deep sea (where only blue-green light penetrates); it is general in insects, present in at least some spiders and crustaceans, but curiously absent from the cephalopod molluscs, whose eyesight in other respects is excellent (Messenger 1981).

Polarisation, the plane in which light waves vibrate, is not a quality of light we can distinguish, because the photopigment molecules in human receptors are aligned randomly in the rods and cones. In insects this is not always the case, as the molecules tend to be arranged along the long axis of the microvilli that carry them. Light waves vibrating with their electrical vectors parallel to the microvilli, and hence the pigment molecules, will preferentially excite appropriately aligned cells, and so an 'analyser' for polarised light should consist of pairs of receptors with their microvilli arranged at right angles. In the 1940s Karl von Frisch discovered the ability of bees to use this sense, in an extension of his studies on sun-compass navigation. When the sun is obscured, bees can still accurately signal the direction of a food source in their dances, provided some blue sky is visible. Skylight is polarised as a result of scattering in the upper atmosphere, and this pattern of polarisation is related predictably to the position of the sun, whether or not this is visible (Wehner 1976). The problem for the bee is to infer the position of the sun from the pattern of polarisation in those parts of the sky that are visible and, although there is still much debate as to how this is done, there is no dispute that bees and many other invertebrates have this capability. Amongst vertebrates the evidence is more equivocal, but it is likely that some fish, amphibians and birds possess this navigational aid to some degree. Polarisation is not confined to sky light. All reflecting surfaces polarise light to some extent, and water surfaces do so completely at certain angles. Similarly, the silvery sides of fish like the herring, which serve as camouflage in ordinary light (Denton 1971), become visible when viewed through polarising glasses. It is surely no coincidence that predatory squid have the microvilli in their retinae arranged in a square array, a perfect analyser to defeat the camouflage.

Exploiting the image

The qualities of light mentioned above—intensity, colour and polarisation—are all properties of small regions of the visual field. It is their disposition within the field as a whole that informs an animal of the locations and identities of objects in its surroundings.

It is not very easy to make a classification of these 'whole image' properties in a way that is useful to students of animal behaviour, but three categories stand out to me as being the most convenient. First, the overall pattern of light and dark. Secondly, the geometrical distribution of light and perhaps colour in small parts of the visual field, corresponding to reasonably small objects. This category, which overlaps with the other two, would include significant patterns like landmarks, sign stimuli and 'releasing' stimuli, Thirdly, patterns of stimulation resulting from movement, either of the animal itself or of objects in the surroundings.

The overall pattern

Protozoa, flatworms, insect larvae and most other 'simple' animals can select their appropriate environment by using their eyes or eyespots to orient towards or away from sources of light. These simple responses form the basis of an important classification of animal orientation (Fraenkel & Gunn 1961). In *orthokinesis* an animal simply slows down where the intensity is appropriate and speeds up elsewhere, thereby spending more time in the right environment. In *klinotaxis* (shown by insect larvae) the animal continuously oscillates its head, and turns more to the side on which it finds the least (usually) or most light. In *tropotaxis* the input from the two eyes is compared, enabling the animal to turn towards or away from light without the sampling movements of klinotaxis. In *menotaxis* an animal may maintain a particular ratio of light intensity falling on the two eyes, which is a simple way of maintaining a constant direction with respect to the sun, and some molluscs use this to guide their foraging paths. None of these mechanisms requires more than the most rudimentary image-forming capabilities from the eyes, and they lead to delightfully simple models for connection of the receptors and the locomotor organs. Problems arise, however, when an animal can detect local differences of intensity within the overall light distribution. Some snails, for example, orient towards dark objects with a larger vertical than horizontal extent. Sawfly larvae migrate across the Australian desert towards vertical objects (which might be trees) and will even follow a moving human if that is all there is to fixate (Meyer-Rochow 1974). Clearly, in these examples, we have something like pattern recognition, the extraction of features with definable properties from the overall pattern. At this point the language of the kinesis/taxis description of behaviour ceases to be useful.

Pattern recognition

To pick out from the visual image certain configurations, and classify them according to their significance for action, means that an animal's nervous system must have within it a set of templates that can be matched by appropriate retinal patterns. Humans, after years of learning, seem to have so complete a set of templates that few everyday objects are seen as unrecognisable. At the other extreme, the sawfly larva may recognise only stationary vertical objects (potential food to which it orients) and moving objects (predators at which it spits). Between these extremes lies a vast and largely unexplored spectrum of pattern recognition capabilities, some of which are learned (a digger wasp learns to recognise the landmarks around its nest; an octopus can be trained to discriminate between a square and a diamond) and some not. In the latter category fall many of the 'sign stimuli' or 'releasers' that are so dear to ethologists. The claw-waving display of fiddler crabs (*Uca*) and the striking poses adopted during mating by jumping spiders are well known (for review see Wehner 1981), and there is a vast literature on the stereotyped postures adopted by birds in sexual and territorial encounters (Fig. 1.4). Even some flowers, especially

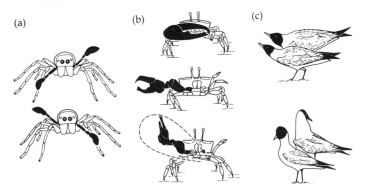

Fig. 1.4. Displays involving both pattern and motion in (a) jumping spiders, (b) fiddler crabs and (c) black-headed gulls. (a and b after Wehner 1981, c after Tinbergen 1959.)

orchids, have developed structures that mimic the form and sometimes the scent of insects, which are lured into attempting copulation. One has the impression, especially with the sexual displays of birds and arthropods, that the signals have evolved to be as striking, precise and unambiguous as possible—as though to give a not very subtle analysing mechanism as good a chance as possible to recognise them. The relative unobtrusiveness of mammalian visual communication, conversely, probably reflects the greater competence of the pattern-recognising machinery.

Patterns of movement

Many displays incorporate movements which are as much a part of the releasing stimulus as is the geometry of the pattern itself. The perception of motion, however, has a much more general role in the life of all moving animals. J.J. Gibson (1950) pointed out that most of an animal's knowledge of the three-dimensional structure of the world comes not from binocular vision but from the way the image flows across the retina during locomotion. For an animal moving in a straight line there is a pattern of velocity on the retina that is zero straight ahead and behind, and maximal to each side (see Fig. 1.5). The nearer to the observer an object is, the faster it is seen to move, and there is a very straightforward relationship between the retinal velocity ($\dot{\theta}$), speed, position and distance:

$$\dot{\theta} = \frac{V \sin \theta}{d},$$

where θ is the direction of an object relative to the body axis, d the object distance and V the animal's speed. If an animal is equipped with velocity-detecting neurons that measure $\dot{\theta}$ at each position θ on the eye, and it has some estimate of its own speed over the ground, then it is a simple matter for the nervous system to translate the outputs of velocity detectors into object distances. Motion information can thus supply the third dimension of the visual world which is missing in the two-dimensional image on the retina. The interesting thing about motion vision is that it provides a large amount of immediately useful information without requiring any recognition of geometrical pattern. For example, an expansion of the motion pattern directly ahead means that an obstacle is approaching, and you must avoid it or land on it. A coherent

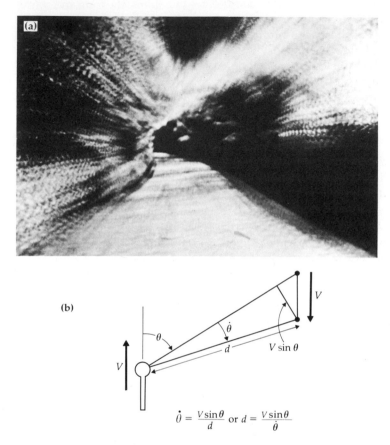

$$\dot{\theta} = \frac{V\sin\theta}{d} \text{ or } d = \frac{V\sin\theta}{\dot{\theta}}$$

Fig. 1.5. (a) Pattern of retinal motion as seen by a moving animal. The picture was made by superimposing 0.5 s of cine film taken from a moving car. The basic pattern radiates from a 'pole' in the direction of motion, and the magnitude of apparent movement depends on the distances of objects and their position in the field of view. (b) The formal relation between retinal motion ($\dot{\theta}$, an animal's speed (V) and its distance from a stationary object (d). If the animal has an estimate of its speed, and the location of the object in the visual field is known from the eye coordinates (θ), then the distances of the object can be obtained directly from its retinal velocity.

rotation of the whole flow field means that you yourself have rotated, and must correct your course. A change in the speed of the image beneath a flying animal means that it has changed either height or ground speed, perhaps as a result of a change in the wind. Perhaps because human vision is dominated by the fovea and the pictorial view of the world it provides, this kind of information has largely been ignored even though we certainly use it (Lee 1980). However, for a bird or an insect moving rapidly through a complicated three-dimensional environment, retinal motion rather than pattern is likely to be the most important single aspect of its vision.

1.2.2 The mechanical senses

Mechanical deformation of sense cells is the basis of a great many distinct sensory modalities. These comprise touch (deformation of the body surface), vibration (substrate-borne oscillation) and sound (air- or water-borne oscillations). Ultrasound—as in bat and cetacean sonar—is almost a separate sense. The lateral line sense of fishes monitors local water displacements, and the same 'hair cell' receptors occur in the semicircular canals (measuring angular rotation), the maculae of the sacculus and utriculus of the inner ear (measuring linear acceleration, especially gravity) and also the organ of Corti in the mammalian cochlea, responsible for hearing (see Fig. 1.6). To this list should be added proprioceptors of various kinds: the muscle spindles of vertebrates and stretch receptors of arthropods, which measure length, and the tendon organs of vertebrates and companiform sensilla of insects, which respectively monitor muscular and cuticular stress. Finally, there is an assortment of minor senses—baroception (depth/pressure), wind reception, and the recently substantiated magnetic senses of insects and birds—that all rely on the detection of mechanical movement. Animals, not surprisingly perhaps, have found all the uses for mechanical transducers that centuries of technology have realised, in the form of the microphone, accelerometer, strain gauge and so on. Of this list, the acoustic senses stand out as distinct because they alone supply information about objects at a distance. Touch and the lateral line sense concern events at or close to the body surface, and the others are all proprioceptive in that they inform an animal of the movements of the parts of its body, or of the results of its own actions. This discussion will be confined to the acoustic senses, although similar principles apply to the other mechanical senses.

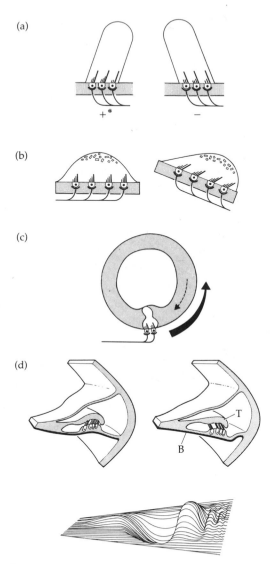

Fig. 1.6. Different types of mechanoreceptor in vertebrates, all based on the same kind of hair cell. (a) Water-movement detectors in the fish lateral line, with the hair cell cilia embedded in a gelatinous cupola. Cells respond positively to displacements towards their kinocilium. (b) Gravity receptor in the inner ear. The cupola contains dense otoliths which shift as the structure tilts. (c) Rotational-acceleration detection in the semicircular canals. The cupola acts as a pendulum in a

1.2.3   *Acoustic senses*

Vibration

A disturbance on the surface of a substrate, whether this is the ground, a water surface or a spider's web, will set up a travelling wave which attenuates with distance. Such propagated waves can be used as an early warning signal of the approach of a potential predator, and in many animals elicit a defensive response. Rather more interesting are the cases where animals use substrate vibrations as a medium for communication. Male wolf spiders, for example, drum on the ground with their palps during sexual displays (Rovner 1967) and fiddler crabs similarly bang the substrate with their enlarged chelae. In at least one case, the bug *Canthophorus* where individuals indulge in vibratory duets, the signals seem to be tuned to the acoustical properties of the substrate, since proper song alternation only occurs if the bugs are joined acoustically by the stems of their host plant (Gogala 1978). Besides communication, for whatever purpose, substrate vibrations can be used to locate prey. Backswimmers (*Notonecta*) and water striders (*Gerris*) can orient accurately towards an insect struggling in the water surface film, using differences in time of arrival of the ripples to determine direction. Similarly, agalenid spiders that spin sheet webs determine prey location from the direction of travel of the wave in the web. Orb web spiders also use vibration for detecting prey, though here of course the task of localisation is simplified by the radial geometry of the web. On the basis of the temporal structure of the vibrations, orb web spiders can distinguish prey from conspecifics. The problem with substrate vibration as a means of localisation and identification is that the transmission properties are not ideal. Soil in particular tends to attenuate oscillations rapidly, especially higher frequencies, and inhomogeneities distort the waves that are transmitted. Some

---

fluid-filled canal, and bends with the fluid as this lags behind the motion of the head. (d) Sound receptors in the cochlea of the ear. The basilar membrane (B) is tuned to different frequencies at different places along its length, producing a wave-form shown (exaggerated) below. These movements are detected by the hair cells of the organ of Corti whose stereocilia are embedded in the gelatinous tectorial membrane (T). (a–c from Schmidt 1978, d from Yost & Nielsen 1977.)

scorpions, nevertheless, manage to locate burrowing prey quite accurately from the vibrations set up in the (relatively) homogeneous sand. Water surfaces have more predictable properties, but only in still weather.

Sound and hearing

In contrast to soil, fluid media (air and water) transmit oscillations with minimal distortion over long distances, in the form of pressure waves. In the immediate vicinity of a vibrating source, an insect's beating wings for example, there may be large displacements of the medium itself which can be detected by organs that are simply bent by bulk movements of the air or water mass. The antennae of mosquitos and the neuromasts of the lateral line system of aquatic vertebrates are examples of this kind of displacement monitor. This 'near-field' effect, however, attenuates very strongly with distance, in contrast to the pressure wave or 'far-field' effect which attenuates less dramatically and which we usually speak of as 'sound'. To hear a sound requires a transducer that reconverts pressure changes into motion, and this generally has the form of a thin membrane: the eardrum of terrestrial vertebrates, a wide variety of tympanic organs in insects, and the gas bladder in some fish.

Sound waves contain information about intensity, frequency and direction. In addition, the way in which intensity and frequency patterns vary over time can be very important. Bird song and human speech, when displayed on a sound spectrograph, both show continuously changing patterns of amplitude and frequency (pitch), with the 'message' coded in all three parameters, time being the third (Fig. 1.7). Although the pitch of a sound is determined physically by the rate of variation of sound pressure with time, hearing organs do not, in general, pass on information about individual pressure oscillations; indeed this would be impossible for, say, a 10 kHz sound when the maximum firing rate of nerve fibres is 1000 cycles per second. Frequency is thus best regarded as a quality of sound, not unlike colour in vision, that is extracted in a different way from the overall modulation of the sound pattern with time.

Sound within the human audible range (20 Hz to 20 kHz) is largely a modality of communication between animals, and although the direction of origin of a single sound may be accurately located, there is no acoustic equivalent of the visual image. (This is not necessarily true for ultrasound, as we shall see.) Some inanimate objects emit sound—water, rustling leaves—but not enough for an acoustic sense to be generally useful as a navigational instru-

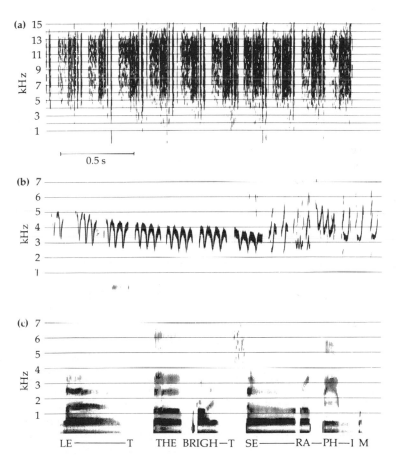

Fig. 1.7. Sonagrams of **(a)** singing insect (meadow grasshopper, *Chorthippus parallelus*), **(b)** bird (redwing, *Turdus iliacus*), and **(c)** human (unaccompanied voice singing the aria 'Let the bright seraphim' from *Samson* by Handel). In (a) most of the information is in the temporal sequence of rather undifferentiated chirps. In (b) the song is strongly modulated but still very stereotyped. In (c) there is both temporal and frequency modulation, but it is now very difficult to 'see' the stereotyped pattern of the song.

ment, except in the case of orienting to another sound-producing animal. The noisiest groups in the animal kingdom are the insects and vertebrates (including fish) plus a few crustaceans like the snapping shrimps. Insects in general do not distinguish different

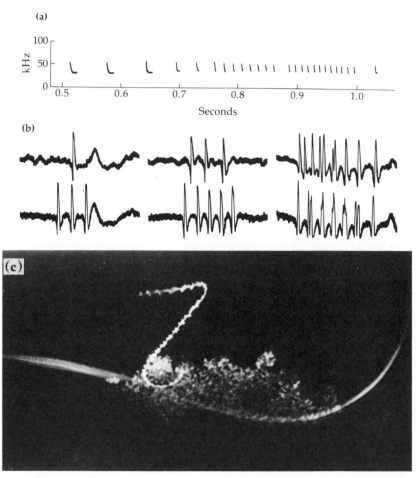

Fig. 1.8. **(a)** The call of a vespertillionid bat approaching a potential prey, with the characteristic 'buzz'. **(b)** The response of the neurons in the two ears of a moth to an approaching bat. Firing rate is greater in the nerve from the ear closer to the bat (below) and increases as the bat gets closer. The two neurons fire about equally as the bat passes overhead (oscillograms on right). (c) Evasive action taken by a moth which soars upwards just as a bat, seen here as a long white streak, swoops by on a capture path. (a from Sales & Pye 1974, b and c from Roeder 1965.)

frequencies, although there is some evidence that bush cricket ears may do so. It is usually the syllabic structure of insect songs that contains the message, which is commonly quite straightforward. Michelsen (1978) describes the import of the field cricket's song: 'I am here, I belong to species X, females are welcome, males are not'. A typical cricket song consists of several pulses of sound at a characteristic frequency in the range 2–8 kHz. Each pulse lasts a few tens of milleseconds with a gap of similar length before the next one. The whole pattern is repeated after a pause of 100 ms or more. The basic frequency acts as a carrier, with the information contained in its modulations. Other insect songs may be more complicated; the record for complexity is probably held by a katydid, *Amblycorpha*, with a song in three distinct phases that lasts 42 seconds (Alexander 1960). In addition to communication at modest carrier frequencies, some insects have ears that respond to the ultrasonic cries of bats, in the range 40–100 kHz. Again, these are 'tone deaf' in that they do not distinguish between frequencies within the range that stimulates them, but they do respond differentially to intensity (bat distance) and direction, and moths equipped with such ears can successfully evade capture by cruising bats, as shown in Fig. 1.8 (Roeder 1965).

Amongst the vertebrates, fish can certainly hear, and many make sounds, but there is little evidence for frequency discrimination above 1 kHz. Below that frequency, receptors in the ear tend to fire synchronously with the peaks and troughs of the disturbance, thus behaving as vibration receptors. Mammals, on the other hand, all possess devices for separating out the frequency spectrum along a resonating structure, the basilar membrane, so that it is the *location* of an active receptor rather than its rate of firing that specifies the frequency of the sound reaching the ear (Fig. 1.6d). This in turn means that neuron firing rate can code intensity, and the ear as a whole can encode the whole frequency–amplitude continuum of complex sounds. This presumably accounts for the extreme versatility of auditory communication in birds and mammals, and contrasts with monotone buzzing in insects. The amphibian system is intermediate between that of fish, where intensity and frequency are not well distinguished, and the bird and mammal arrangement, where they are. Frogs have three populations of auditory neurons, covering low-, mid- and high-frequency ranges. Of particular interest is the observation that in different species of frog the response of the ear is tuned to the major frequency components of the call of that species; this no doubt helps species recognition to some degree (Capranica 1976).

There is accumulating evidence that the nature of animal sound is tailored in various ways to the manner in which sound propagates in the environment (Michelsen 1978; Gerhardt 1983). In forests, for example, there seems to be a 'transmission channel' at around 2 kHz within which sound will travel twice as far as it will at frequencies of 500 Hz or 3 kHz. Forest-dwelling birds tend to use pure tones around 2 kHz, whereas grassland birds, in different acoustic conditions, use a much wider range of sound frequencies in their songs. In the sea, low frequencies attenuate very little, especially where particular temperature conditions create a layer in which sound is effectively trapped, and under these conditions it is possible, at least in theory, for the low notes in whale songs to travel for hundreds of kilometres. 'Ecological acoustics' promises to be an intriguing subject.

Ultrasound

As the frequency of a sound (f) increases, its wavelength (λ) decreases. They are related by $\lambda = c/f$, where c is the velocity of sound ($340\ ms^{-1}$ in air; $1400\ ms^{-1}$ in water). The significance of this is that at high frequencies the lengths of sound waves become comparable with the sizes of the objects they encounter (1 cm at 34 kHz compared with 1 m at 340 Hz). As in microscopy, objects larger than a wavelength will reflect waves, those comparable to a wavelength will cause scatter to some degree, and those much smaller will barely perturb the waveform and so be undetectable. By raising the frequency of sound emission into the range where the wavelength is a centimetre or less, some animals have managed to create an acoustic image-forming system good enough not only to navigate by but also to enable them to catch small flying prey. In short, ultrasound, like light, behaves as rays and provided the detecting system is good enough it may be used in the same sorts of ways. The enormous bonus is that it can be used at night.

The three groups of animals that exploit this difficult but rewarding modality are nocturnal birds (two species), cetaceans and, above all, bats (Sales & Pye 1974; Busnel & Fish 1980). The microchiropteran bats emit calls in the ultrasonic range, 40 kHz up to about 120 kHz. Some sweep through a range of frequencies, whilst others, notably the horseshoe bats (Rhinolophidae), tend to maintain a constant high frequency throughout the call. The pulses are brief, a few milleseconds or less, and are repeated every 10–100 ms, speeding up into a characteristic 'buzz' just before an attempt at prey capture (see Fig. 1.8a). The prey usually consists of nocturnal insects like mosquitos and moths, but a few bats will take frogs and even fish. Basically, a bat needs to extract from the echo of its call the target's range and position, and enough information about the 'quality' of the target to distinguish prey from inanimate objects. Range is straightforward: it is the speed of sound multiplied by half the echo time, or 1.7 m per 10 ms. The bearing of the target in the horizontal plane can be extracted from differences in time of arrival at the two ears (these differences are a matter of *microseconds*) or from intensity differences. These are the same clues that we use to locate a sound within our audible range. In addition, some bats waggle their ears during sound emission, no doubt scanning the environment for echoes and building up an acoustic picture in the manner of a line scan on a TV screen. How direction in the vertical plane is extracted is still largely a mystery (as it is for humans) though it is likely that the acoustical properties of the large pinna play an important role. There is no doubt that bats can distinguish different properties of objects in their path from the frequency spectrum of the reflected echo. Small objects preferentially reflect higher frequencies, and the surface texture of the target also modifies the frequency qualities of the echo.

1.2.4   Electric senses

It is logical to consider electrolocation by fish after bat echolocation because these are the two 'active' senses, in which animals emit the energy they subsequently receive. A few deep-water fishes light up the water around them with luminescent photophores (McCosker 1977) but all other senses rely on energy or chemicals provided by the environment. The demonstration that certain fish (the African mormyrids and South American gymnotids) use electric pulses for navigation and communication is quite recent. In 1951, Lissmann found that the mormyrid *Gymnarchus* produces a continuous electrical signal from its tail with a frequency of about 300 Hz and an amplitude of 30 mV. This is far too weak to be used to stun prey, unlike the much more powerful discharges of the electric ray *Torpedo* or the eel *Electrophorus*. *Gymnarchus* could avoid obstacles in the dark and when swimming backwards. In common with other weakly electric fish, *Gymnarchus* has a peculiar method of locomotion, using only its dorsal fin and keeping the body straight; the significance of this is that the animal keeps a constant relation between the electric field produced by its tail and the electroreceptors on the head. Interestingly, Lissmann also found that the fish would vigorously attack an electrode that played back its own discharges into the water. It thus seems that, as well as navigational cues, the electric discharges provide the fish with a means of communicating with other electric fish. The electric sense probably evolved as a substitute for vision in murky water. The field set up by the tail (Fig. 1.9) is detected by receptors on the head that are derived from cells of the lateral line system. Objects in the near surroundings distort the field, conductors drawing the field in and non-conductors expanding it, so that the

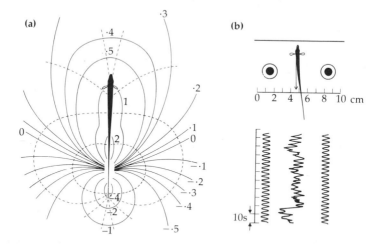

Fig. 1.9. (a) The voltage field around an electric fish (*Eigenmannia*) at the peak of its electric organ discharge. The figures are peak voltages in mV. Stippled lines show the flow of current. **(b)** The 'electromotor response'. Plexiglass rods (15.6 mm in diameter) moved invisibly within porous pots induce a corresponding motion in an electric fish. Using its electric sense, the fish attempts to keep station with respect to the objects in its environment. (Both from Heiligenberg 1977.)

pattern of field strength the animal's head receives, when suitably decoded, provides an electric image of the surroundings. This image is good enough for *Gymnarchus* to distinguish between a porous pot containing a 2 mm glass rod, and an empty pot. This corresponds to a voltage difference at the head of the order of 1 μV cm^{-1}. In spite of the system's exquisite sensitivity, the electric sense can be used for navigation only at short distances, less than about a metre. It would be surprising if electric emission were not also used for communication, and there is now considerable evidence that discharge patterns are modified during aggressive and sexual encounters (Hopkins 1977). It seems likely too that individual fish can recognise each other from the shape of the waveform of single discharges.

Besides the actively emitting species, there are other fish that have no electric organs but which can detect the presence of prey animals by using electroreceptors (the ampulli of Lorenzini) to detect their muscle potentials. Kalmijn (1971) found that sharks and rays could detect live buried flatfish, but not if the fish were cut up, ruling out chemoreception, and not if they were placed in an insulating box. The sharks would, however, attack a pair of buried electrodes. It seems then that the electric sense is very versatile in water, providing a passive sense for detecting prey, a communication channel rather like sound, and a substitute for short-range vision with features in common with ultrasound.

1.2.5 Chemical senses

The detection and identification of chemicals in the environment is a faculty all animals possess. In human physiology, chemical senses are usually divided into taste and smell, defined respectively as the detection of water-soluble chemicals already in the mouth and of airborne molecules emanating from a distance and detected via the nose. The distinction is one that blurs slightly as one looks at other animal groups—for a fish all chemicals are water-borne, for example—but it is still useful to preserve a distinction between chemical senses that require contact and those that act over a distance. The former are principally concerned with determining the palatability of food, and the latter with detecting potential food, with communication of various kinds mediated by pheromones, and in a few cases with establishing the identity of particular places, like the home streams to which salmon return for spawning.

In humans it is difficult to know quite how to sort out the chemical patterns that make up particular tastes or smells into components that make sense functionally or physiologically. A sensation of sourness results from acid, and H^+ ions are presumably the stimulating component. On the other hand, glucose, D-leucine and beryllium chloride all taste sweet, whereas quinine, L-leucine and magnesium sulphate all taste bitter. Plant alkaloids, which are often toxic, tend to have the lowest thresholds (8 μmol.l^{-1} for quinine compared with about 0.01 mol.l^{-1} for sucrose), suggesting a protective role for taste as well as a more general food-quality monitoring function. Human smell is just as difficult to compartmentalise, and although a list of seven 'primary odours' has been defined, these do not seem to correspond to the electrophysiological properties of single olfactory neurons, which are only weakly specific. It may be that the whole pattern of stimulation of the olfactory bulb is important in establishing a smell's identity—a sort of odour image (Gleeson 1978). In contrast, insect chemoreceptors tend to be rather specific. In the blowfly *Phormia* each of the four classes of labellar receptor responds to a range of substances that does not overlap with those stimulating the others (Dethier 1974), and at a behavioural level many insects show extreme specificity in terms of the species of vegetation they will eat.

In comparison with obviously directional senses, like vision and hearing, olfaction gives a rather poor indication of the location of the source of a stimulus. In air or water an odour will diffuse out from the source, and a very small amount of turbulence will destroy a concentration gradient that might assist an animal to home in on the origin. (Most people are familiar with the problem of trying to locate a bad smell by 'klinotaxis'.) Nevertheless, there are some situations where odour can provide a more accurate directional cue. In a light wind a source will produce an odour plume downwind, and in good conditions this may be tens of metres long. A male moth, for example, that encounters a few molecules of female attractant—and there is a good evidence that single molecules of the pheromone are adequate to stimulate single hairs (Schneider 1974)—needs only to fly upwind to find the female, or at worst to improve its chances of doing so. Another way of increasing the vectorial precision of a smell is to lay it on the ground, as in the trail pheromone of the fire ant (Fig. 1.10; Wilson 1963). The volatility of the chemical is important, as a stable substance would persist for too long and tracks would become indecipherable, and the ant pheromone in fact lasts for only a few minutes. In an interesting comparison, Wilson suggests that the amount of information about source direction and distance contained in the trail of an ant is not very different from that in the waggle dance of a bee, a predominantly visual and auditory display (von Frisch 1962). Pheromones in insect societies, and in fish and some mammals amongst vertebrates (Todd 1971; Shorey 1977), look like being at least as important as any of the other modalities as vehicles for communication, although at present the difficulties of establishing the nature of the minute quantities of chemicals involved are formidable.

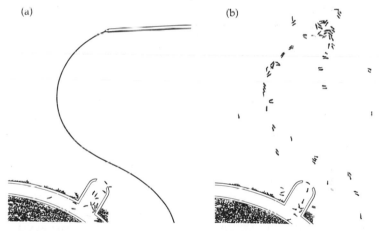

Fig. 1.10. The spatial information contained in the pheromone trail of a fire ant (*Solenopsis saevissima*). In (a) is shown the line along which pheromone from the Dufour's gland of an ant has been smeared with a stick. Some time later (b) many ants have been attracted from the nest and are concentrated along this line or are milling about in a confused way at its further end where the trail suddenly ceases without leading to anything. (After Wilson 1963.)

1.3   Conclusions

In this brief review we have had to ignore many important senses. Temperature, pressure, humidity, touch and the currently

popular magnetic sense are all omitted as being perhaps peripheral
to the main theme of the chapter. This has been the interplay,
within the major senses, of the ways in which the two main
features of any stimulus—its identity and its position—are
extracted. Throughout the history of ethology, this distinction has
been made: an innate releasing mechanism indicates identity and,
classically, releases an action pattern (fixed or otherwise), but that
action pattern must be directed properly. Lorenz, evading the
constraint of the word 'fixed', wrote of the 'taxis' component of
the action pattern. In neurophysiology too the distinction between
pattern and position has often been made. As we have seen, vision
is supremely valuable for locating objects and is good for identifying them provided that the brain is up to the pattern-recognition
task. At the other extreme, olfaction is the most problematic posi-
tional sense, but can be used to identify plant foods or animal
pheromones with great precision and with very little skill required
on the part of the CNS. Ultrasound and electroreception have
taken on some of the attributes of vision where that modality is not
available, though judging from the amount of brain space that is
devoted to these modalities in bats and electric fish, at very considerable cost. Audible sound is perhaps the finest medium for
accurate communication between individuals, though it is not of
much value in dealing with inanimate objects, and is only really
exploited in the insects and land vertebrates.

If there are phylogenetic comparisons to be made between the
sensory abilities of different animal groups they would be along
the lines of increasing specificity and precision in the arthropods,
and increasing flexibility in the vertebrates and perhaps the
cephalopod molluscs. Insects seem to see only what they need to
see, and to learn only when they need to learn. Mammals, on the
other hand, learn to distinguish not only those patterns that must
be identified for immediate survival, but also those that might be
valuable in the future. Marler (1961) pointed out that the sensory
equipment an animal possesses may be thought of as a filter
which only passes the information the animal needs for its main
chores—feeding and reproduction—whilst excluding the vast
irrelevant residuum. For an insect larva choosing a host plant, or
an adult insect selecting a mate, this filter may be situated as
peripherally as the sensory membranes of the chemoreceptors
themselves. In frogs, some degree of visual object recognition may

occur at the level of the retinal ganglion cells (third-order receptor neurons), and in mammals it is unlikely that anything that usefully resembles a filter is present before the cortical representation of the visual fields (fifth- to seventh-order afferent neurons), and even then only 'multi-purpose' features seem to be extracted, lines and corners for example, rather than those of direct biological importance. If the idea of a sensory filter is still a useful concept, what one can say is that it retreats inwards as the number of discriminations an animal is called upon to make increases.

The luxury of possessing rather general-purpose senses like our own is no doubt related to size and longevity. A small, short-lived animal needs neurally parsimonious equipment that will work accurately from the first minute of life. It is salutary to remind ourselves that mayflies emerge with no mouthparts, indeed there are a few species that do not even have legs. They are in no position to learn how to mate.

1.4   Selected reading

There is, at present, no single book that covers the subject of *comparative* sensory physiology at the level of an advanced university course. I would therefore suggest a combination of a book on human senses, such as Schmidt's *Fundamentals of Sensory Physiology* (1978), and the chapters on the senses in Prosser's *Comparative Animal Physiology* (third edition, 1973). For more detailed information there is the monumental *Handbook of Sensory Physiology* edited by Autrum and others, now running to 12 volumes.

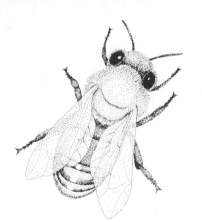

CHAPTER 2
SENSORY GUIDANCE OF
MOTOR BEHAVIOUR

T.S. COLLETT

2.1   Introduction

Animals move around and, to do so effectively, they need information about both their environment and their relation to it.
Indeed as the last chapter stressed, sensory systems are primarily
designed not for constructing a picture of the outside world but for
guiding activity within it. That this point should need any
emphasis is a reflection of the hierarchical way our brains work.
When we look around, whether we ourselves are stationary or
moving, we are aware of a fixed external world, in which objects
that we recognise as having a certain range of possible uses occupy
a defined location in space. This apparently effortless understanding of our surroundings is assembled out of a mixture of glimpses,
memories and expectations. But all we notice is the end product,
and this we can use to plan our movements within the world.
However, when we come to perform actions like picking up a cup
or approaching a doorway, we are largely unconscious of the
sensory input from our muscles, joints, and vestibular and visual
sysems that enables us to execute the movements properly. The
actions are under automatic control; we only see in the end
whether we have been successful in executing them.

This chapter continues the theme of the previous one and
explores from two points of view how sensory information guides
actions. First, there is the mechanistic question of the way in which
sensory and motor systems interact. Secondly, there is the question of what information animals extract from the world in order to
control their behaviour. The environment is full of signs as to what
is happening in it, and animals read and interpret these in various
ways and with various degrees of sophistication according to their
habitat and the complexity of their behaviour. The problems

animals encounter and the various solutions they have adopted will be illustrated in a series of case histories.

2.2　Chemoreception in bacteria

An organism's purposive-looking actions need not imply that it has an internal representation of the environment which it uses to construct plans to achieve its goals. Some animals do work in this way, but many just guide their movements using a limited number of cues from their surroundings. Behaviour that looks complex often results from very simple rules operating within a complex environment. The trick in this case is to choose the right cues and the right rules. The bacterium *Escherichia coli* provides a nice example of how natural selection has operated to enable the organism to pick out the appropriate signals for exploring its liquid environment for sources of food (Koshland 1979).

A bacterium's habitat is so foreign to our own intuitions that it is necessary to spell out what it feels like to be an organism that is only 2–3 μm long and 1 μm wide. Being so small, inertia is insignificant and the only forces a bacterium experiences are those of viscosity. Scaled up to our size, it is as though we had to move through liquid asphalt. One peculiarity of living in a situation in which viscous forces are vastly greater than inertial ones is that you cannot swim away from your immediate surroundings, you only shed them slowly. Thus while a whale can swim through water scooping up plankton as it goes, a bacterium can do no better than to wait for individual nutrient molecules to reach it by diffusion (Purcell 1977). The reason why bacteria swim is to find places where there is an abundance of nutrient molecules in the surrounding liquid. To do so they monitor how the concentration of relevant molecules in the immediate neighbourhood changes over time by measuring the occupancy of chemical receptors on their body surface.

The way in which diffusing molecules behave means there is little advantage in covering more than a very small fraction of the body surface with receptors. Once a molecule has reached the cell it tends to stay in the vicinity, bumping into the cell repeatedly so that it has many chances of finding a receptor. Consequently, even on something as small as *E. coli* there is no room for many classes of receptor and there are known to be in total about 30 distinct types,

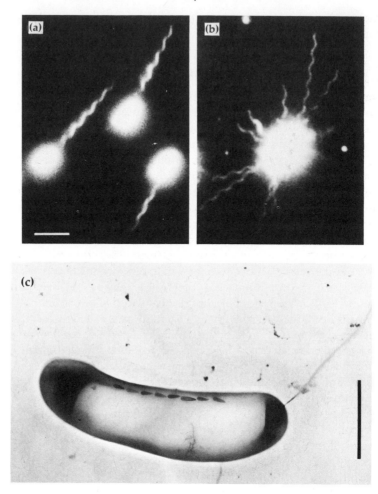

Fig. 2.1. Flagella and magnets in bacteria. (a) *Salmonella typhimurium* during forward swimming when the flagella form into a bundle. (b) Flagella dispersed as they are during tumbling. (c) South-seeking bacterium from New Zealand. Chain of arrow-shaped magnetosomes is parallel to long axis of cell. Pointed shape is atypical. Calibration bars 5 μm in a and b, 1 μm in c. (a and b from McNab & Ornston 1977, c from Blakemore *et al.* 1980.)

making the bacterium responsive to a wide range of attractants and repellents (Berg & Purcell 1977).

Perceived changes in the concentration of attractants and repellents must then control the organism's overall direction. Many common bacteria swim by rotating stiff, helical flagella which are formed into a bundle to the rear (Fig. 2.1a). A bacterium changes direction by momentarily reversing the direction of rotation, when the bundle disperses and the flagella stick out akimbo (Fig. 2.1b). This causes the cell to turn by a random amount and it then sets off in another direction until the motor briefly reverses again. Its course thus consists of a series of straight segments interrupted by tumbles which alter its direction unpredictably (Berg 1975). The only parameter of movement under sensory control is how often tumbling occurs. The bacterium drops its tumbling rate in order to keep going in the same direction when things are improving and the encountered concentration of nutrients rises. But it does the reverse in order to sample many different directions when the concentration of a repellent increases. Changes in tumbling frequency thus bias its random walk towards attractants and away from repellents (Koshland 1979). There is nothing directional in either sensory input or motor output. But given the properties of its surroundings—a decaying fish will advertise itself by molecules leaking from it—the bacterium's machinery has sufficient structure to guide it effectively within its world.

There still needs to be careful matching of input and output parameters, however. Since the bacterium does not know its orientation with respect to the external world, it is important that the straight segments of its swimming are adjusted to be long enough for the bacterium to assess during a single run how the concentration is changing. Information concerning what happened earlier is of little use after a tumble has resulted in a change of direction.

2.3   Magnetic bacteria

Because bacteria have so little mass, gravity has a negligible effect upon them in comparison with the random buffetings they receive from water molecules. Thus, should bacteria need to distinguish up from down, they must find a cue other than gravity by which to do so. It has recently been discovered that some bacteria exploit the earth's magnetic field for this purpose (Blakemore & Frankel 1981). Away from the equator, the earth's field has a vertical component, and at the poles a freely suspended magnetic compass points vertically. Bottom-living bacteria which are adapted to low oxygen tensions use the vertical component of the earth's field to keep themselves headed downwards and away from dangerously high oxygen levels.

Inside the bacterial cell is a chain of small (c. 5 nm) beads of magnetite (Fig. 2.1c) on the dipole of which the earth's magnetic field produces a torque which is sufficient to steer bacteria against the Brownian motion of water molecules. The form of the magnet is cunningly designed to maximise its magnetic dipole. The size of each bead is just right for it to form a single magnetic domain, while their arrangement in a chain means that the dipoles of all the beads will be aligned parallel to its length. In species with flagella at one end, the dipole is oriented within the cell to point it tail upwards, causing the bacterium to swim downwards. This is the case in both the northern and southern hemispheres, which means that the dipoles within the cell must be oriented in opposite directions in the two hemispheres (Blakemore & Frankel 1981).

Larger multicellular organisms, like bees and birds, which also respond to magnetic fields must have sense organs to detect the magnetic field and then use sensory information from these to steer themselves by. It is only in bacteria that biological magnets are relatively so large that the earth's magnetic field can act directly to turn the whole organism.

2.4   Control of swimming direction in paramecia

Even without a nervous system it is possible to have sensory and motor mechanisms that are to some degree spatially organised. Paramecia are single-celled aquatic organisms about two orders of magnitude larger than bacteria. They are covered with motile cilia which beat in a complex pattern to propel them in a helical path. When necessary, the ciliary beat can be reversed and the paramecium swims backwards. Thus should a paramecium collide with an obstacle, a mechanoreceptor is stimulated at the anterior end and the animal backs away. However, if it is bumped from behind a mechanoreceptor at the rear is stimulated causing the animal to accelerate forwards (Eckert & Randall 1978).

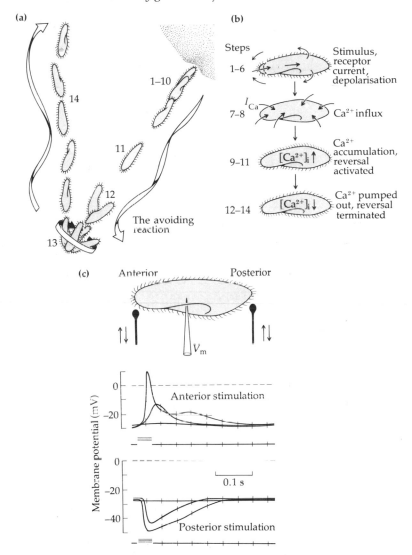

Fig. 2.2. The avoiding reaction of *Paramecium*. **(a)** After hitting an object the *Paramecium* backs off (steps 1 to 13), turns and heads off in a new direction (13 and 14). **(b)** The flow of calcium ions associated with this sequence of behaviour. **(c)** Receptor potentials recorded across the cell membrane which are induced by anterior or posterior stimulation. Thick lines show voltage changes resulting from stimulation at three different intensities. Lower trace shows duration and relative intensity of stimulus. (From Eckert & Randall 1978.)

How are inputs and outputs linked? In multicellular animals information from individual receptors is carried by nerve fibres and after appropriate processing can be routed to selected effectors causing a directed response. In paramecia internal signalling is by means of the potential difference across the membrane of the cell and this is 'read' by all the cilia. In common with most cells, the inside is at a negative potential with respect to the outside. Stimulation of the front mechanoreceptor causes the potential across the membrane to drop (Fig. 2.2) and this signal spreads over the whole cell. Voltage-sensitive channels, like those in nerve fibres, are distributed locally round each cilium and respond to the fall in potential by opening and letting calcium flow down its concentration gradient into the cell. The calcium then interacts with components in the cilium to reverse the beat, before being vigorously pumped from the cell. Stimulation of the rear end activates a different species of mechanoreceptor which locally increases potassium permeability. This raises the membrane potential and accelerates the beat. Thus the spatial position of a mechanoreceptor on the body determines the sign of the change in potential it induces, and this in turn tells cilia whether to beat forwards or backwards (Eckert *et al.* 1976). Although paramecia can respond in a directed way to stimuli in their environment, the flexibility of control is limited at all levels: the information available from the sense organs, the means of linking receptors to effectors, and the restricted range of behaviour which results from the concerted action of the cilia.

2.5   Reading the retinal image: insect orientation

The special properties of light make vision the most accurate source of spatial information about the world (see Chapter 1). Larger animals living where light is abundant generally possess well-developed eyes and place reliance on vision for understanding their surroundings and for guiding their movements within these. But to exploit the vast potential offered by a retinal image animals must have a nervous system which is equal to the job of making sense of the complex patterns of light and shade which pass across their eyes. Particular patterns on the retina must be matched to events in the outside world and studies of both animal and machine vision have emphasised what a complex task this is.

For instance, any image on a two-dimensional retina could be the projection of one of a multitude of possible three-dimensional configurations but, by making assumptions about the real world and the behaviour of objects in it, our visual system is usually able to single out one of these as the most likely.

One unconscious assumption we make which helps constrain our interpretations is that most objects are rigid. Thus when people are shown two spots moving coherently on the surface of a screen, the spots are vividly seen as the ends of a rigid rod. Should the spots also be made to move relative to each other, the rod is never seen as changing its length, but always as a rigid object describing a trajectory in depth (Johansson 1975). It is only when a moving image cannot be interpreted to fit the rigidity rule that the object is seen to change shape. In addition to requiring rules for interpretation, animals also need them for linking visual and motor patterns. Both classes of rules are well illustrated by flying insects, many of which have beautifully developed eyes with intriguing specialisation for particular tasks (Fig. 2.3).

Insects of a number of species chase other insects, either as potential mates or as potential prey. The initial recognition process, that is the selection of an object as a suitable target, is often very crude, a target being anything that is small, moves within a certain range of velocities, and contrasts darkly with the background. Size and speed are as measured on the retina, so that a fast, distant bird evokes the same response as a nearby conspecific or a pebble tossed into the air. The response to the target is also governed very simply by spatial properties of the retinal image which can be transformed straightforwardly into motor commands. Chasing consists of turning to face the target and moving towards it. The more peripherally placed the target image is on the retina, the faster the insect turns (Land & Collett 1974; Reichardt & Poggio 1976). Retinal position is thus translated into angular velocity. To control their distance from the target some flies use the size of the target image to govern their forward velocity. The image is expected to be of a particular size. If the image is larger than it should be, the insect knows it has approached too close and decelerates. If the image is too small, the insect flies faster to catch up (Collett & Land 1975).

Similar information, the position and velocity of the image of the target, is used somewhat differently by the males of some

Fig. 2.3. Sexual dimorphism in the St. Mark's fly (*Bibio marci*). The male (above) is blessed with an 'extra' dorsal eye which it uses to spot and follow females (below) seen against the sky. (Courtesy of J. Zeil.)

species of hoverfly to decide in which direction they must head in order to intercept a passing conspecific (Collett & Land 1978). In the absence of any prior information, the computation of an interception course seems quite a formidable task. One needs to know, for instance, the direction and speed of the target in space. But, by making certain assumptions about the behaviour of the target it wants to catch and by limiting its interception strategy to those targets, the fly has been able to reduce the problem to manageable proportions. Flies of a single species are sufficiently uniform in size that a hovering male can first expect to spot an approaching conspecific at a particular distance. If now the male also assumes that the conspecific has a constant cruising speed, then the velocity at which the fly's image travels across the retina when it is first visible will tell the male the target's direction of movement. When the image velocity is zero, the target should be heading straight for the male, when image velocity is equal to the ratio of the cruising speed to the sighting distance, then the target's direction is perpendicular (Fig. 2.4). The male is thus furnished with a simple rule for translating the position and velocity of the image on its retina into an appropriate interception course.

As insects fly around they need to avoid obstacles, to decide when to land, to find their way back to a particular spot, to keep on course and to maintain their orientation in space. Vision can help with each of these tasks and in all cases insects have formulated the problem so that they can use a relatively raw retinal image to guide their behaviour.

Several elementary cues provide a reliable index to an insect's orientation in space. Light generally comes from above, so that to control roll an insect should turn to keep the dorsal part of its retina maximally illuminated, and many species do this. This reflex is mediated in part by the compound eyes and in part by the ocelli, which are simple eyes with unusual optical properties, designed specifically for such a job. There are generally three ocelli set between the compound eyes, two facing laterally and one forwards. They have wide fields of view, each arranged so that in normal flight one half looks at the ground, the other at the sky. The lateral ocelli monitor roll, the medial one, pitch (Stange & Howard 1979).

In most eyes, the retina and lens are carefully adjusted to keep the image focused in the plane of the receptors. The ocellus is a striking exception to this principle (Wilson 1978). It consists of a thick lens lying in front of a cup-shaped receptor layer. The receptors are situated in front of the focal plane of the lens, so that the image in the receptor plane is blurred. To match the low optical resolution there is massive neural convergence, with the thousand or so receptors feeding on to 25 fibres which transmit information to the central nervous system. The ocellus thus actively discards spatial resolution. On the other hand it can provide a rapid assessment of the overall distribution of light and dark which is

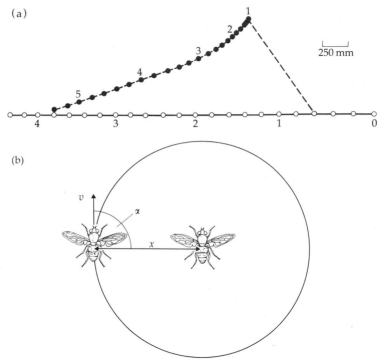

Fig. 2.4. Interception flights in the hoverfly, *Eristalis* sp. (a) Male chasing a pea blown towards it from a pea-shooter. Position of pea (indicated by open circles) and fly (filled circles) are shown every 20 ms. Dashed lines show when male is likely to have first sighted the pea. When the male detects a target, it does not head directly towards it but computes an interception course. (b) Method male uses to compute the direction of target's approach (α). Hovering male detects target which comes within range (x). If female travels directly towards male, angular velocity of target on retina ($\dot{\theta}$) is zero. If target flies at a tangent to circle, $\dot{\theta}$ is v/x, where v is the target's velocity. For intermediate directions $\dot{\theta} = (v \sin \alpha)/x$. Provided male can make reasonable assumptions about the values of v and x and measure $\dot{\theta}$ it is able to infer α and so decide on an interception course. (From Collett & Land 1978.)

just what is needed to control an insect's position with respect to the earth and sky.

The orientation of edges in the world gives another cue to spatial orientation. In open country the horizon is the predominant feature. Flying locusts expect their environment to contain horizontals, and orient themselves appropriately (Goodman 1965). On the other hand the retina of insects flying close to the ground will be filled predominantly with the vertical edges generated by plants. Flies, it seems, assume their environment to be full of verticals and, when suspended in front of a striped pattern, align themselves so that the dorsoventral axis of their eyes is parallel to the orientation of the stripes (Srinivasan 1977).

The distribution of image motion as an animal moves is a rich source of information which is cunningly exploited by insects to guide their flight. A nice example is the way houseflies decide when to land (Wagner 1982). When a fly approaches a stationary object at a uniform velocity, the size of the image of the object increases. But since in this case the fly has no *a priori* reason to expect the object to be of a particular size, the image size on its own will not tell the fly its distance from the object. The rate at which the image size changes is also by itself not a useful cue, since the rate at which it does so depends not only on the distance of the object from the insect, but also on its size and the insect's velocity. However, the ratio of the rate of expansion of the image to its size does contribute valuable information. It tells the fly when it can expect to collide with the object, were the fly to continue at the same velocity. Films of flies landing on spheres of different sizes have shown that they begin to decelerate when the ratio reaches a fixed value. Thus it is not the fly's distance from an object that governs its decision to land but its 'time to contact' (Lee 1974).

Insects are by no means the only animals to exploit bulk image flow and it is intriguing that when we detect image movement on the periphery of the retina we generally feel that it is us and not the world that is moving, in contrast to image movement restricted to the fovea, which we see as being generated by the movement of external objects (Brandt *et al.* 1973).

In the examples considered so far there has been no need to assume that insects have any particular knowledge of a specific environment. The behaviour that occurs in a given situation can be accounted for by rules, which are applied generally, operating on

certain features of the environment. However, many insects
return repeatedly to the same spot, often guided by visual landmarks (Wehner 1981). When a hive is moved a small distance from
its customary position, returning bees will hang in a cloud around
the place where the hive used to be, presumably drawn there by
landmarks in the neighbourhood. Feats of this kind imply that
insects must retain some kind of representation of the immediate
vicinity. But what form does this take? Training experiments on
ants and bees suggest that the insect does not construct a Cartesian
map of its surroundings, but that it learns the appearance of the
landmarks as viewed from the nest or foraging spot (Wehner 1981;
Cartwright & Collett 1982). It is as though it takes a panoramic
snapshot from that position. The insect's direction of movement
when trying to regain the spot is then guided by comparing this
snapshot with the image on its retina. To give an oversimplified
example, suppose that a bee has learnt to forage at a point midway
between two landmarks. Should it find itself too close to one of
them, then the retinal angle subtended by that landmark will be
larger than expected. This difference will induce a command tending to push the insect away from the landmark. The retinal size of
the other landmark is smaller than it should be, pulling the insect
towards it. The resultant of these two effects is to move the bee
towards the midpoint. A computer model using principles of this
kind behaves much as real bees do (Cartwright & Collett 1982).

Although descriptions of the kind emphasised in this section
can account for certain aspects of sensorimotor behaviour, they
neglect many subtleties. For instance, it is the custom of males of
many species of fly to hang around in crowds, but despite the
general hubbub a chasing male usually restricts its attention to just
one area of its visual field, following no more than a single target at
a time and somehow excluding the other possible targets from
contributing to its tracking response (Wolf & Heisenberg 1980).
Individual animals commonly show flexibility in the way they take
notice of different aspects of their world at different times. A
hungry toad orients to worm-like objects. A satiated toad ignores
them, and physiological experiments suggest that the relevant
visual neurons in its optic tectum may then be silent (Ewert 1980).
Foraging bumblebees fly from flower to flower looking for a particular species and ignoring others from which their sisters, or they
at another time, are happy to forage (Heinrich 1979).

Similarly, the model of landmark guidance sketched out above seems somewhat limited in scope when one contemplates the extraordinary spatial memories of some insects. Several genera of bee are known to follow stereotyped foraging routes, going from plant to plant in a fixed order. The record is held by South American orchid bees with 'trap lines' which they inspect daily and which may extend over 20 km (Wehner 1981).

This section has stressed how projections of the world on to a two-dimensional retina can provide information for guiding actions. One interesting feature is that the linking rules between retina and behaviour are sufficiently simple that it does not strain one's credulity to imagine how they might be converted into neural hardware Another feature is the fluency with which insects are able to move within three-dimensional space despite the two-dimensional nature of the input to their rules. This is not, of course, to say that insects do not measure depth. Mantids for instance probably use binocular cues to judge whether their prey is within range (Rossel 1980), and locusts seem to employ motion parallax to assess how far they must leap to land on a twig. Before jumping they perform specialised side-to-side head movements and the consequent image motion tells them the distance of the twig (Wallace 1959). However, in such cases the insects' small size and the low resolving power of their eyes limit to at best a very few centimetres the range over which they can measure depth.

2.6   Maps and three-dimensional representations

It is to vertebrates that one must turn for good examples of animals which plan their actions using a three-dimensional map of their immediate environment. One prerequisite for constructing such a map is an ability to measure distances between objects in the external world. Comparative psychologists have in general studied such perceptual problems in relation to what are called the 'constancies', that is whether an animal can judge the real sizes of objects at different distances and, more rarely, how accurately depth intervals can be measured at different distances. Many animals have been shown to exhibit size constancy and rather fewer, depth constancy. In other words, over a certain range of distances they are aware of the physical sizes of objects and the separation between them.

The more complete and accurate a representation an animal has of the outside world, the greater is its potential ability to formulate plans and the greater the potential flexibility of its behaviour. It is only if the spatial layout of an animal's surroundings is known with some precision that a route to a new destination can be planned in advance. Thus the bee cannot use the information contained in its snapshot of an array of landmarks to guide itself anywhere except back to the spot it has learnt. Other animals make more versatile maps. Gobiid fish (*Bathygobius soporator*) have the problem of moving between scattered pools at low tide when they cannot see from one pool to another. But at high tide they can swim over the whole area and the information they obtain from such a vantage point seems to be used to direct their path from pool to pool (Aronson 1971).

People form a precise three-dimensional representation of the arrangement of objects in their close vicinity and can use it to guide their path. To demonstrate the accuracy of their map, subjects were blindfolded after surveying an unfamiliar obstacle course from the starting position. Lack of vision in no way prevented them from planning and taking an appropriate route through the obstacles. Subjects showed how detailed their internal floor plan was by neatly avoiding each obstacle, as long as the route was covered within the seven seconds that it takes for the information in the map to decay (Thomson 1980).

The possession of size and depth constancy, however, does not necessarily mean that an animal can negotiate a complex obstacle course and that it has an internal map of its surroundings in the same sense that we do. Toads, for instance, are aware of the real sizes of things and this allows them to judge whether a moving object should be eaten or is so large that it is better avoided (Ewert 1980). They also have some idea of how far apart obstacles are, and so know whether a gap between obstacles is big enough to afford them passage, or whether a hole in the ground is so shallow that it can safely be stepped into (Lock & Collett 1979, 1980). Although they can make such judgements about individual features of their surroundings, they seem unable to assemble the information into a 'map' which will allow them to reject certain routes in favour of others. They are forced to follow the dictates of simple visuomotor rules and obeying these can easily lead them into blind alleys (Collett 1982).

Vision is not the only modality through which a map can be constructed. Bats use echolocation to form a picture of places they frequent (Neuweiler & Möhres 1967), and blind people can also build up a map of their surroundings. For example, after a congenitally blind two-and-a-half year old had been guided along some of the possible paths between four objects, she had evidently built up a map that was sufficiently accurate to let her take the direct path when asked to go from one object to another, even though it was a path new to her (Landau *et al.* 1981).

It is far from clear how such maps are actually coded in the nervous system and how the information in them is read out to guide actions. Introspection suggests that when we close our eyes and move around we have an internal image of the layout of objects seen from our current vantage point. If, for instance, we approach an object with closed eyes, the object is 'seen' to loom up. This suggests that the image not only guides movements, but that the movements are fed back to change the format of the image so that it matches the expected visual input. Under normal circumstances with vision available, one's sensory expectations are confirmed (or not) and the internal image continually revised. In familiar surroundings, expectations of what should be there are often of more importance in guiding an animal's path than what is actually there. Bats frequently bump into obstacles when they occur unexpectedly in a well-known place (Schnitzler & Henson 1980) and may attempt to land on a favourite toe-hold which an unkind experimenter has removed (Griffin 1958). Rats and fighting fish are known to follow an accustomed detour well after the removal of the obstacle which originally made the detour necessary (von Uexküll 1957).

2.7 Ecological constraints on perceptuomotor behaviour

An animal's peripheral sensory equipment is very obviously shaped by its ecological niche. The Mexican blind cave fish (*Anoptichthys jordani*) has abandoned vision and indeed has no eyes, relying instead on its lateral line organs to detect water movements. When it swims it generates a bow wave and the fish seems to pick up a great deal about its environment from the way nearby objects distort this wave. Perceptuomotor behaviour is also moulded by habitat, and equipped with the same sense organs different animals notice different features of their surroundings.

Many species have an innate disposition to avoid a cliff edge, and if placed on a 'visual cliff' (a sheet of clear glass covering a steep drop) will keep clear of the drop (Walk 1978). But to an aquatic animal suspended in water, a submerged cliff edge is no danger and may be disregarded. Thus aquatic turtles, unlike terrestrial ones are unperturbed by cliffs. Similar species differences are shown by gull nestlings placed in a nest on a raised platform. When scared by the approach of a potential predator, species which nest on beaches or sandhills are more likely to jump off the platform than those nesting on cliffs, which tend to remain motionless. In some birds such species-typical behaviour is seen after a variety of rearing conditions, whereas in others it can be modified by an appropriate upbringing. Species of terrestrial and arboreal mice also behave according to their natural habitat when placed on narrow platforms, and these differences are retained after generations of identical laboratory rearing (see review by Walk (1978)).

Without an independent measure of a species' ability to assess depth, it is difficult to be certain whether an animal which is fearless on cliff edges is just blind to them or whether its sensorimotor rules are modified so that cliffs are treated appropriately. The latter is certainly the case for ducklings of some species which normally nest in holes in trees and must as very young nestlings leap from their nesthole to reach water and find food below. Such nestlings are untroubled by the steep side of the cliff, and when hatched in an incubator chose at random between the vertical and shallow sides. However, they could clearly discriminate between the two. They often ran from the shallow side and jumped off the steep one as though they were leaving their tree. Similarly, mallard ducklings (*Anas platyrhynchos*) would push off from the 'edge' of a visual cliff, as if launching themselves into a pond (Kear 1967).

Animals which live in trees and clamber among branches face special problems. Branches provide their highways, but not all are equally good. Some branches are culs-de-sac. Some are so thin and pliable that they change direction when walked upon. Some that are thin and dead will not bear an animal's weight. Chameleons, as soon as they emerge from the egg, are able to cope with the problems presented by such an environment (Harkness 1977, personal communication). They like to climb to the tops of things and when faced with a vertical maze constructed out of artificial branches, they will clamber up and at choice points avoid arms with blind endings. They are also wary of thin branches which may bend and lead them to an unanticipated destination. It is intriguing that jumping spiders which hunt on plants show equal facility in taking the right route to avoid dead ends (Fig. 2.5).

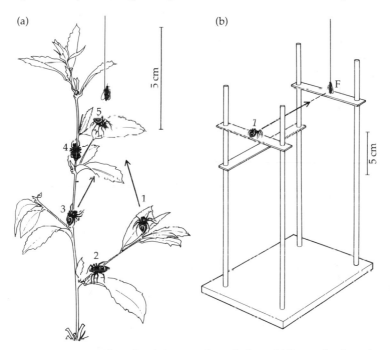

Fig. 2.5. Jumping spiders plan their routes through plants. (a) Route taken by male *Phidippis pulcherimus* to reach a fly. As soon as the spider oriented towards the fly, the latter was removed so that the spider made the whole of its ascent without view of its goal. (b) Artificial plant. Spider starting at 1 first turns towards fly (F) and then almost without mistake orients directly towards the 'bridge' and crosses it to reach its prey. (From Hill 1979.)

2.8   Neural models of the external world

A central theme of this chapter so far has been the way in which information about the external world is translated into a form suitable for steering actions. It is now time to consider more directly what is known about the neural machinery involved. Two aspects of this question intensively researched over the past decade will be summarised here as an example of how neuro-physiologists approach these problems: first, the representation of spatial information within the vertebrate brain, and secondly the use of spatial information to guide eye and limb movements.

Micro-electrode recordings have revealed that, by and large, individual neurons within sensory areas of the brain perform very specific jobs and respond to a narrow range of stimuli. In some areas of the mammalian auditory cortex, for example, single cells will discharge a burst of nerve impulses to sounds falling within a narrow frequency range. Different cells respond to different frequencies and to cover the whole range to which the ear is sensitive an array of neurons is needed. Such arrays tend to be spatially ordered, with the frequencies to which single cells respond shifting in a graded manner across the surface of a cortical area (Merzenich *et al.* 1977). Orderly maps* of this kind are common in the nervous system and have presumably arisen for a number of different reasons. Thus there is a topographic represen-tation of the retina on to the surface of the striate cortex, in which single cells respond to visual targets within a restricted region of space and in which neighbouring areas of cortex map neighbour-ing regions of visual space. The cortex seems designed to detect local properties of the retinal image such as the orientation of edges, the direction of movement, binocular disparity and so forth. The best way to perform the necessary local computations is to retain the neighbourhood relations of the retina (Barlow 1981).

Both of the previous examples are cases in which sensory surfaces on the body, the cochlea and the retina, are mapped out spatially on to a central neural structure, and there may well be developmental reasons why central projections should be organised in this way. However, an exciting discovery made over the past few years is that similar forms of spatial coding are used to represent features of the external world which cannot be read directly from the surface of peripheral sense organs, but whose assemblage requires considerable central processing. There exists

*The term 'map' is used very differently by neurophysiologists and by ethologists and psychologists, so differently that using it both ways in this chapter should not lead to confusion.

within the owl's midbrain, for instance, a very precisely ordered map of the direction of sound with respect to the animal's head, similar to maps of visual space. In this map, single neurons respond to sounds coming from a small area of space, with different neurons listening to different regions (Fig. 2.6). Unlike the eye, where the position of a stimulus on the retina defines its horizontal

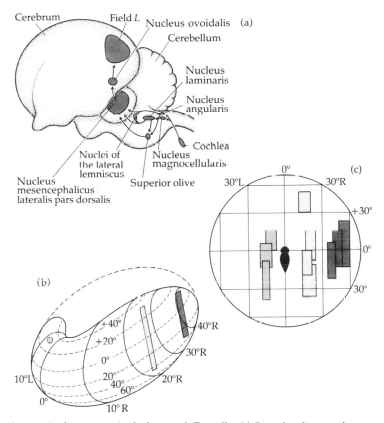

Fig. 2.6. Auditory maps in the barn owl, *Tyto alba*. (a) Central auditory pathway from the cochlea to the forebrain. Much processing occurs in the midbrain, where the auditory map is to be found in the mesencephalicus lateralis pars dorsalis (MLD). (b) Enlarged view of MLD showing the map. Neurons in this structure respond only to sounds coming from a particular region of space in front of the owl, as shown in (c). About half of the MLD is devoted to the region of space within 15° of the mid-line, where the owl locates sounds most accurately. More of the map is concerned with space below the bird than it is with space above, as suits an animal which swoops down on its prey. (From Knudsen 1981.)

and vertical coordinates, the direction of a sound is determined by the owl, as by mammals, by comparing the times of arrival and intensities of sounds at the two ears. It is the result of such binaural comparisons that go to make up the map (Knudsen 1981).

Another very elegant example is provided by the auditory cortex of insectivorous bats. In keeping with the animal's reliance

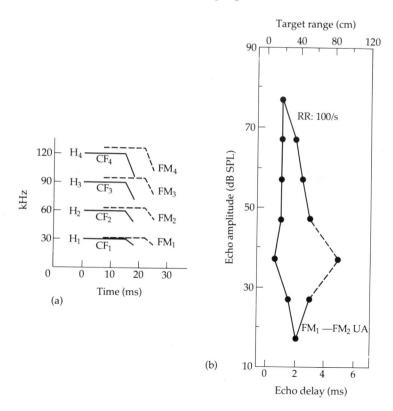

Fig. 2.7. Delay-sensitive neurons in the moustached bat. (a) Schematic diagram of bat's orientation call (solid lines) and returning echo (dashed lines). The four harmonics (H$_{1-4}$) each contain a long component at constant frequency (CF) followed by a short component during which the frequency drops (FM). Echo delay is the time difference between corresponding components of call and echo. Only FM component gives unambiguous timing information. (b) Relation between echo delay and intensity of echo needed to evoke a response from a range-sensitive neuron. Neuron only responds for delays and intensities which lie within the 'diamond'. The best response is when echo delay is 2 ms. Note that neuron only fires for echoes which are neither too loud nor too soft. (From Suga et al. 1981.)

on ultrasound for investigating its environment, the auditory cortex is relatively very large compared with that of other mammals. It is subdivided into several physiologically distinct areas, each of which provides information about some feature of the bat's insect prey (Suga *et al.* 1981). Separate regions are concerned with the distance, size, relative velocity and wing-beat frequency of the prey. Thus, bats compute prey distance with astonishing precision from the delay between emission of the orientation sound and time of arrival of the returning echo reflected from the insect. In one area of the auditory cortex, single neurons hardly respond to either the outgoing sound or the echo alone, but fire strongly to pairs of such sounds with a particular delay between them (Fig. 2.7). Neurons sensitive to the same delay are grouped together and the delay producing maximal response alters progressively along the cortex from a fraction of a millisecond to about 18 ms, equivalent to target distances ranging from just under 10 cm to 3 m. There are few neurons for distances over 2 m, which fits nicely with the finding that this is the distance at which several species of bat first seem to notice small obstacles.

A final and dramatic example is the multimodal representation of space within the principal midbrain visual area known as the optic tectum in lower vertebrates and the superior colliculus in mammals. A mouse can determine the direction of an object by seeing it, hearing it or feeling it, the last through vibrissae on its snout, and in the colliculus there are maps of space in different layers for each of these modalities which are arranged roughly in register (Dräger & Hubel 1975).

Similarly, the tectum of certain snakes (pit vipers, like the rattlesnake, and boid snakes, such as the python) harbours both a map of visual space and one obtained through special sense organs sensitive to infrared radiation. These 'pit organs' enable snakes to catch warm-blooded prey in the dark. They act like crude pin-hole cameras and produce an inverted image of the outside world on the heat-sensitive membrane at the bottom of the pit. Rattlesnakes have a single flask-shaped pit situated below and in front of each eye, with a field of view somewhat narrower than the eye's. Pythons, on the other hand, have several pit organs arranged in a row, each monitoring a few tens of degrees of space. In the latter case there is the intriguing problem of ordering the row of inverted images to produce the coherent map of space which is found in register with the visual map in the tectum. The fact that snakes should succeed in arranging their neural connections in the complicated way needed to do this stresses again that these internal spatial maps of the outside world must have a crucial role in the control of behaviour (Newman & Hartline 1982).

The behaviour of cells in the rhesus monkey's superior colliculus suggests that the important property coded by the firing of collicular cells is indeed the location of a stimulus, rather than its quality. Whereas cells in the striate cortex (Fig. 2.8) respond only to stimuli of the correct orientation, disparity and so forth, those in monkey colliculus are far less selective and just require that the target should move and be located in the correct region of space (Wurtz & Albano 1980). The same distinction is found in the auditory region of the owl's midbrain. Neurons forming the auditory spatial map respond to a wide range of frequencies, in contrast to other non-spatially organised areas in which single neurons are sensitive only to a limited frequency band (Knudsen &

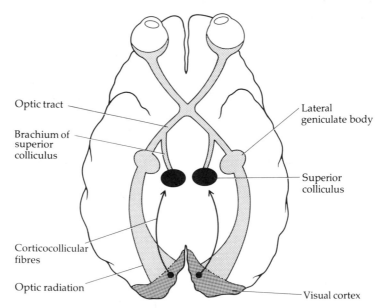

Fig. 2.8. Highly simplified sketch of the primate visual pathway. Axons from the retina go to the lateral geniculate nucleus in the thalamus or to the superior colliculus. The lateral geniculate projects to the striate or primary visual cortex at the back of the brain. (From Angevine & Cotman 1981.)

Konishi 1978). Furthermore, monkeys or humans with striate lesions can direct their eyes and limbs towards visual targets (presumably using the direct retinal input to the colliculus, see Fig. 2.8), but cannot identify the target's shape (Humphrey 1974; Pöppel *et al.* 1973). Indeed, people suffering from cortical blindness are quite unaware when tested that they have the capacity to locate visual stimuli by looking or pointing, and deny that they see anything at all.

The heartening thing about sensory maps of this kind is that the parameters coded in them should mostly be so accessible to our intuitions. This suggests there may be some truth to the generalisation that sensory areas of the brain are segregated into subregions, each of which directly reflects a behaviourally significant dimension of the outside world.

2.9 From sensory maps to motor codes

It has often been argued that, at some high level in the brain, actions are coded not as a sequence of detailed commands to muscles, but as movements specified in spatial and temporal co-ordinates, so that at this level the form of the motor code is compatible with the way in which the external environment is represented within the brain. Perhaps the most direct evidence for this view comes from highly practised motor skills like handwriting. If people are asked to write the same word or letter at different speeds, in different sizes or on vertical as opposed to horizontal surfaces, the writing looks the same, even though different muscles will contribute to the movements of the pen when letter size or orientation of the writing surface is changed. Measurements of the time intervals between forming different features of a letter or a word have shown that the ratio of the lengths of neighbouring intervals remains constant despite wide variations in their absolute durations (Viviani & Terzuolo 1980). This study provides quantitative evidence that a given spatio-temporal pattern can be generated in a variety of ways, implying that the pattern must be represented in a more versatile form than a sequence of specific signals to individual muscles.

Other lines of research suggest more directly that movement codes are designed to mesh with sensory maps. Sensorimotor skills are learnt more easily and performed with fewer errors, the greater the spatial correspondence between sensory input and motor output (Fitts & Posner 1973). The aiming of a finger at a target light is a motor skill in which the spatial coordinates of stimulus and response are highly compatible. For this task and others like it, performance in humans and rhesus monkeys is not slowed by an increase in the uncertainty of the position or the timing of the target light (Fitts & Posner 1973; Georgopoulos *et al.* 1981). This is just the kind of result that would be generated by a system in which spatial information is processed in parallel through many channels within congruent sensory and motor maps.

The question then arises of how a translation is made from what is perhaps a spatial representation of a movement to detailed commands to muscles. Designers of robots have had to face the same problem and have found it a difficult one. Mechanical arms commonly consist of several linked segments. Complete instructions to the arm must specify the position and trajectory of each segment. However, for the kinds of tasks manipulators have to accomplish, the most important feature of their behaviour will usually be the movements of the most distal segment (or hand). Controlling devices external to the arm will only be interested in what the hand is up to and will issue command signals in terms of hand movements. There is thus a need to translate from what the hand is asked to do to full instructions for all the segments. The difficulty of doing this increases with the number of segments and their degree of freedom. The solutions which have been adopted vary from simplifying approximations to the use of 'look-up' tables containing the translation. It has proved simpler to generate such tables from a record of the manipulator's previous performance than to derive them analytically (see, for example, Winston & Brown 1979). Real limbs, it is proposed, avoid some of these difficulties by restricting the available degrees of freedom: a vast number of movements which our limbs could in principle perform are never realised (Bernstein 1967; Greene 1972). However, the problem of how spatial commands are turned into muscle twitches is far from being solved.

At a less detailed level, the way primates change their direction of regard illustrates some of the mechanisms which enable the same action to be achieved in different ways. To turn and look at an object we use some combination of head and eye movements. Sometimes we turn both head and eyes and sometimes we keep the head still (Fig. 2.9). Does the central command to redirect gaze need to specify exactly how the movement is distributed between the two effectors, or can it just produce a signal proportional to the new direction of gaze and leave the details of implementation to other control systems? The question was answered, in the monkey, by recording commands to the two effectors. Initially the eye muscles receive the same signal whether or not the head moves. Take the case where, at the end of the movement, the whole change can be accounted for by the altered position of the head. Neck and eye muscles receive command signals that are almost simultaneous. The eye, having lower inertia, starts to move first and the head follows more slowly. As it does so, the eyes rotate back in their orbit, such that when the movement of the head

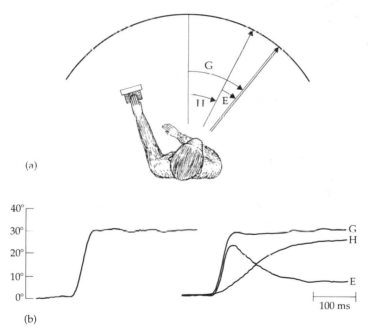

(a)

(b)

Fig. 2.9. (a) Coordination of eye and head movement in primates. Direction of gaze (G) is a function of the position of the head with respect to external coordinates (H) and the eye with respect to the head (E), such that G = E+H. (b) Direction of gaze can be changed either by swivelling the eyes (left) or by using a combination of eye and head movement (right). In the latter instance the eyes move first and the head follows more slowly. Time is given along the abscissa, amplitude of movement along the ordinate. (a from Whittington *et al.* 1981, b from Morasso *et al.* 1973.)

is completed the eyes have returned to their original attitude with respect to the head. Should the head be suddenly and unexpectedly prevented from moving, then the whole change in direction of gaze is achieved by the eyes. What happens is that the same gaze command is available to head and eye, but a peripheral reflex, the vestibulo-ocular reflex, measures the speed and duration of the head movement and tells the eyes to move in the opposite direction, thus automatically subtracting the amount the head moves from the command signal to the eyes. A higher command system need therefore not know in detail how its instructions are apportioned between head and eye (Bizzi *et al.* 1971). Other systems can decide whether the monkey should keep its head still when stealing a surreptitious glance at a dominant male.

2.10   The generation of saccades

One of the few cases in which one can follow, at least in part, events from activity in a sensory map to the production of a movement dependent on that map is the generation of rapid eye movements in primates. This section is concerned with how information from a visual map in the colliculus is translated into a form suitable for guiding eye movements.

When we explore the details of a scene to investigate an object viewed out of the corner of an eye, rapid eye movements direct the fovea accurately from point to point within the external world. These saccadic (jump-like) movements are so quick that there is no time for them to be controlled by visual feedback, so that the location of the target on the retina must be transformed into commands suitable for moving the eyes to a new position.

It is simple to imagine how this might be done in an animal like an insect which in order to place a target on its fovea moves its eyes by rotating its whole body. In this case the position of the target on the retina can be uniquely associated with a turn of a fixed size. Until recently it was thought that visually guided saccadic eye movements in primates were similar. The assumption was that when a target appeared 10° to the left of the fovea a command was issued to move the eyes 10° to the left. The position of the eye in the orbit never had to be explicitly calculated; it was just updated with each saccade. However, it now seems that this hypothesis is mistaken, and that eye movements resemble other spatially directed limb movements (Robinson 1981).

When an animal with mobile eyes wishes to point a finger at a peripheral target, it must work out the position of its eyes in its head and of its head with respect to its body, as well as the position of the target on its retina. Humans can undoubtedly do this: we are able to aim a hammer with an accuracy of a fraction of a degree at a target briefly glimpsed in the dark with our eyes positioned eccentrically in the orbit (Hansen & Skavenski 1977). We compute where the target is with respect to our body and then direct a limb to that location. In terms of eye movements this means that the saccadic command must specify the final position of the eye in the orbit and so must take into account both the retinal position of the target (the retinal error) and the initial orbital position of the eyes. How and where is this accomplished?

Lesion studies imply that visual information for guiding saccades is routed through two parallel pathways, one involving the superior colliculus, the other an area of cortex known as the frontal eye fields. Provided that one of these pathways is intact, monkeys can still make accurate, visually driven saccades (Schiller *et al.* 1980). Visual space in the colliculus is still mapped out with the fovea represented at a fixed point on the map, whatever the position of the eyes with respect to head. Such a coordinate system, specified with respect to the retina, is described as retinotopic and is also found in striate cortex and, in very imprecise form, in the frontal eye fields.

2.10.1 Motor maps in the colliculus

Although the ultimate command is framed in terms of eye *position*, visual inputs arriving at the colliculus interact with what is essentially a map of eye *movements*. Electrical point stimulation of the monkey colliculus evokes saccadic eye movements of short (20 ms) latency. The amplitude and direction of these movements conform nicely to the retinotopic map of visual space: an electrode placed in a given site elicits a saccade of an amplitude and direction appropriate to place a target seen by that region of the map on to the fovea (Fig. 2.10b). The characteristics of the saccade are unaffected by the initial position of the eyes, and should the stimulating current be prolonged, the eyes produce a staircase of saccades of

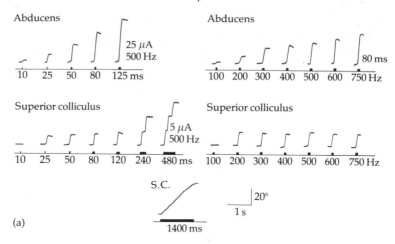

(a)

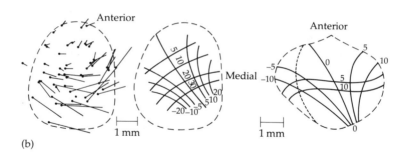

(b)

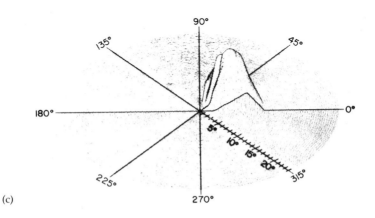

(c)

constant size (Fig. 2.10a). This is the result to be expected from stimulating a spatially mapped sensory area and is also exactly what happens if a target image is clamped in a fixed position on the retina. In both cases the saccadic system receives a constant retinal error signal.

The colliculus is, however, much more than a sensory area and has a distinct and anatomically separate motor map in its deeper layers, which like the sensory map above is organised retino-topically (Wurtz & Albano 1980). The cells in it fire before saccades, but for each cell this is only before saccades made in a particular direction and of a particular amplitude. These parameters of the saccade define the movement fields of individual cells, and in a given area of colliculus the movement fields are roughly coincident with but larger than the receptive fields of visual neurons in the superficial layers (Fig. 2.10c). In some mammals (e.g. cats) orienting movements of the head (Harris 1980) and pinna (Stein & Clamann 1981) can also be induced by electrical stimulation of the colliculus. Broadly speaking, maps of eye, head and ear movements are in register.

A very important feature of the eye-movement map is that its activity is to some extent independent of the visual map in the

---

Fig. 2.10. The motor map in the monkey superior colliculus. (a) A comparison of the eye movements evoked by stimulating either the superior colliculus or an oculomotor nucleus (the abducens). This nucleus contains the cell bodies of the motorneurons innervating the extraocular muscle moving the eye laterally. When the length of the stimulus is increased (left) an electrode in the superior colliculus evokes one, two or three saccades of the same amplitude. However, in the abducens nucleus the size of the movement increases with duration. Similarly, an increase in the frequency of a 0.5 ms pulse is associated with an increase in the amplitude of the movement when the electrode is in the abducens nucleus, but not if it is in the colliculus. Thus in the colliculus it is only the location of the electrode which influences the size of the evoked saccade, in contrast to the abducens nucleus in which signals are frequency coded. (b) Variation in saccade amplitude with position of electrode. Left: plan of left colliculus showing electrode position (•) and saccades resulting from stimulation; the amplitude and direction of the movement are indicated by the length and orientation of the associated line. Middle: saccade map obtained from many such observations. Right: visual map. (c) Three-dimensional representation of the movement field of a cell in the superior colliculus. This is plotted by recording the activity of the cell for different saccade amplitudes and directions. Amplitude and direction shown by position on horizontal plane, size of neural response indicated by the height of peak. (a from Schiller & Stryker 1972, b from Robinson 1972 and Cynader & Berman 1972, c from Sparks *et al.* 1976.)

layers above (Wurtz & Albano 1980). The movement-related cells fire when saccades of the correct amplitude and direction are made in the dark and during the rapid, saccade-like re-centring movements induced when the vestibular apparatus is stimulated by prolonged head rotation in one direction. The motor map thus receives non-visual inputs and these must be adjusted to the map's retinotopic coordinates. This is not a simple task. Consider, as a somewhat hypothetical example, the behaviour of cells in the collicular motor map during saccades made in response to auditory targets. Auditory spatial signals relying on binaural cues are defined with respect to the head, which means that they can specify directly the desired position of the eyes in the orbit. But to make use of the collicular map, the auditory command signals must be made compatible with it, which means subtracting the current orbital position from that specified by the sound. If the colliculus is considered to be representative of the way in which sensory and motor maps are organised, then it would seem that a given motor map is structured to conform to the sensory one to which it is primarily related, and that the transformations needed to construct appropriate output signals are achieved further along the pathway.

2.10.2   From motor maps to motoneurons

Commands thus leave the colliculus in retinotopic coordinates and are translated into head-centred ones outside it. The signals have to alter in several other ways as well before they are in a form suitable for driving the eyes. The eyeball is held in the orbit by a set of six extraocular muscles. These and the surrounding connective tissue constitute an elastic suspension tending to keep the eyes looking straight ahead. For the eyes to point in any other direction, the extraocular muscles which move them must provide a steady force to resist the restoring action of the elastic suspension. Furthermore, to move the eyeball rapidly to a new position in the orbit, the muscles must supply a pulse of force to overcome the viscous drag of the orbital tissues (Robinson 1970).

The muscles thus need to receive both static and dynamic commands. These two components can be seen very clearly in the activity of individual oculomotoneurons innervating them. The static component shows itself as a linear relation between a neuron's firing rate and the position of the eyes, so that when the eyes move from one steadily held position to another the firing rate alters in a step-wise manner. This copes with the elastic resistance. In addition, during the movement there is a high-frequency burst of activity from motoneurons innervating the agonist, which generates the extra force needed to produce the high angular velocities achieved during saccades (Robinson 1970).

It is worth noting that the eye is not a special case, for the head and limbs may also be considered as being suspended in a cradle of damped springs (muscle and connective tissue). To specify the position of limbs, command signals must fix the tensions, and hence the lengths, of the relevant agonist and antagonist muscles (Bizzi *et al.* 1976; Polit & Bizzi 1979).

The behaviour of oculomotoneurons shows not only that there needs to be a shift from eye- to head-centred coordinates, but also that the form of neural coding must alter. In the colliculus, retinal position is represented in a spatial code in which position is signalled by the identity of the active cell within an array. To control the position of the eye, however, the oculomotoneurons use a frequency code in which position depends on the firing frequency of individual neurons and the number of neurons active. Work is just beginning on how this spatio-temporal transformation might be accomplished (Wurtz & Albano 1980).

One reason for setting out these partially solved problems of maps and signal transformations is to make ethologically inclined readers aware that quite complicated behavioural issues are now amenable to detailed neurophysiological modelling, and that by looking inside the brain at how external events and motor actions are represented, one can gain insights into the organisation of behaviour.

2.11   Representing the world

It is almost a truism that to a greater or lesser extent organisms, from bacteria to man, represent features of the external world in the machinery controlling their behaviour. Organisms which rely on simple procedures generally have peripheral or central detectors that classify objects and events in the environment. *E. coli*, for example, not only has very specific chemical receptors which distinguish one sugar from another but it also has a mechanism for classifying sterically different sugars, which are each recognised by their own receptors, as belonging to the same category of molecule, so that they can substitute for one another in attracting a bacterium's approach (Koshland 1979).

It has been stressed throughout the chapter that sensorimotor rules incorporate implicit knowledge about the behaviour of objects in the world. Because in its general features (apart from human interference) the world is a reasonably stable place with abiding laws, the basic structure of many sensorimotor rules can be hard-wired within the brain. Nonetheless, slight adjustments are necessary to compensate for internal changes caused by growth or injury, and even the most automatic reflexes need mechanisms to keep them accurately calibrated.

The vestibulo-ocular reflex, for example, must work very precisely. For good vision it is essential that our eyes receive a stable retinal image, and this means that the eyes must be able to remain stationary in space when the head turns. The vestibulo-ocular reflex helps achieve this by generating accurate compensatory eye movements of the same speed as, but in the opposite direction to, any head movement. The reflex operates without the help of immediate feedback: head velocity measured by the semi-circular canals drives the eyes by way of a very short central pathway. Although in the short term the reflex is an open loop, it has recently been shown that in the long term the reflex uses visual information to adjust itself and improve image stabilisation should its normal functioning be interfered with (Gonshor & Melvill Jones 1971; Miles & Lisberger 1981).

The adaptive capabilities of the reflex have been demonstrated by fitting adult primates with various kinds of optical devices. If monkeys wear spectacles which reduce the size of their retinal image then, for a given head movement, the image on the retina will move less than before. The reflex will now make the eyes travel too far for the image motion caused by head rotation and such overcompensation will generate its own unwanted image slip. After a few days of wearing such spectacles the monkey has adapted to the new situation, and when tested in the dark the gain of its reflex (eye speed/head speed) has dropped from its normal value close to unity to one more suitable for the changed circumstances. If the spectacles are then removed, the gain of the reflex climbs slowly back to its original level. Should the spectacles be reintroduced, the gain will fall with exactly the same time-course as before, indicating that these adaptive changes are as machine-like as the reflex itself (Miles & Lisberger 1981).

One can, as a very first step, imagine an animal like a frog to consist of a bundle of parallel procedures, all mediated by hard-wired neural machinery, with each procedure responsible for dealing with a particular aspect of the world. Thus frogs step round vertical barriers, duck under horizontal ones, and jump away from looming, potentially dangerous objects (Ingle 1976). However, it is rare that an animal's surroundings are so empty that there is no more than a single relevant feature; usually several will

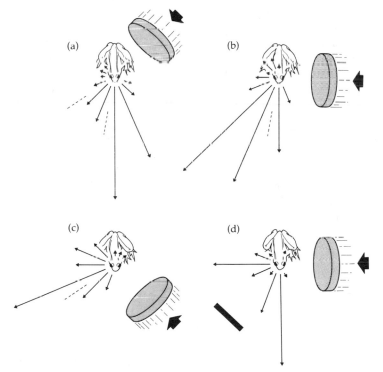

Fig. 2.11. Direction of avoidance jumps in the frog *Rana pipiens* when threatened by an approaching object. Length of spokes radiating from frog's head represents the distribution of jump directions. (a) to (c) show that the preferred jump direction is a compromise between moving forwards and moving directly away from the position of the threatening object. In (d) the frog alters direction to avoid a barrier placed in the path of its normally preferred jump direction. (From Ingle 1976.)

be present simultaneously, as when a stationary barrier in a frog's way causes it to adjust its direction of escape from an approaching projectile (Fig. 2.11).

Animals must therefore be equipped with rules to cope with combinations of features. These must implicitly incorporate laws about the *behaviour* of objects in the outside world, such as: 'predators won't wait for you to finish what you are doing', or 'this is scarce, don't give it up to attend to anything less crucial'. An analysis of the behaviour and of the underlying neuronal circuitry of the marine mollusc *Pleurobranchaea* suggests that rules of this kind are incorporated into the inhibitory and excitatory connections between cells involved in controlling different activities (Kovac & Davis 1980 a,b). The presentation at the same instant of several stimuli which together could potentially evoke simultaneously a number of activities, such as escape, feeding, mating, egg laying and withdrawal, generally does not do so, but results in the performance of just one of them. The choice is not made randomly, but rather it is determined by a series of priorities which seem to make functional sense (see Chapter 3).

As we have seen, many animals are able to learn specific features of their environment, making it possible, for instance, to return at will to a preferred feeding ground. It seems plausible that a large part of the neural substrate required for an animal to understand its particular surroundings makes use of machinery already assembled to mediate sensorimotor rules which operate on general features of the environment. Such rules can be very elaborate and involve large, ordered arrays of neurons, as when actions must be directed to particular regions of space.

2.12   Selected reading

Aspects of bacterial behaviour are covered in a book by Koshland (1980). Wehner's (1981) monumental review explores the whole subject of spatial vision in insects. The vast topic of mammalian visual orientation and the control of eye movements is critically reviewed by Howard (1982).

CHAPTER 3
THE ORGANISATION OF
MOTOR PATTERNS

MARIAN STAMP DAWKINS

3.1   Introduction

In one sense, behaviour is nothing but movement—the movement of whole animals in space, brought about by the movement of limbs, heads and bodies. Even to begin asking questions about behaviour, then, involves classifying and describing movement in some way, putting it into manageable bundles so that we can ask how it happens. The question 'What causes it?' does not make sense unless we have first described what 'it' is. Do we mean the movement of whole herds of animals or the tail flick of a single individual? Imagine trying to describe the migration of a wildebeest herd, in terms of the separate limb movements of all the individuals: clearly impossible. On the other hand, if we were interested in the mechanism controlling walking of a single animal, it might actually be very appropriate to describe in detail what each limb does in relation to the others and indeed to look at the movements of individual parts of separate limbs. So we are up against a problem right at the outset. Behaviour is movement and we wish to investigate its causal basis. But how can we begin to explain that movement, when our subject matter appears to consist of everything from muscle twitches to migrations? If we concentrate on a small part of this, are we just making arbitrary categories for our own convenience? Or are there genuine discontinuities, sections of behaviour which stand out as units for the ethologist in the same way that, say, species do for the taxonomist, or bones do for the anatomist?

One purpose of this chapter will be to try to give an answer to these questions. This will, however, be merely a first step, for the very important reason that it is only part of the truth to say that behaviour is the same as movement. In another sense, behaviour is much more. The movements we see are simply the outward and visible signs of highly complex 'programs' or sequences of instructions to the muscles. Each time a water snail scrapes its radula over a bit of alga, 25 different pairs of muscles contract in particular combinations and sequences (Kater & Rowell 1973). Each time a male mallard courts a female or a lion stalks its prey, the 'symphony beneath the skin' involves hundreds of muscles, coordinated in such a way that each contracts at the right time. When we investigate the mechanisms underlying sequences of behaviour, it is the score for this symphony or the programs of the nervous system that we are seeking and much of this chapter will be concerned with the search for them.

Sometimes the programs of the nervous system are highly specific instructions to individual muscles. They can be called 'low level' programs and they are analogous to the circuit diagrams of a computer. Physiologists are now able to probe the nervous systems of some organisms and find the neural bases of these circuit diagrams. They can analyse them neuron by neuron, a process which Bullock (1976) has called 'circuit breaking'. But sometimes the programs of the nervous system are rather different from this. As the two previous chapters have stressed, animals do not just move around the world in set paths. They negotiate obstacles which may be quite novel to them; they evade predators which are themselves adopting different strategies all the time; they alter their feeding or fighting tactics according to the circumstances of the moment. Much behaviour clearly could not be the result of a rigid series of instructions to the muscles to contract in particular orders at particular times. Indeed, to think of behaviour in this way would be very misleading. A chicken confined in a cage, for example, within sight of its flockmates feeding from a pile of food, will go through a series of actions, but the series will be constantly varied. It may flutter, pace backwards and forwards, push its head under the wire and so on. What is important here is not the exact sequence of movements but the fact that the bird has a goal—it is 'trying' to get at the food. We might well hope to understand how the muscles contracted to produce certain components of the behaviour, such as 'walking' or 'pecking'. But in order to be able to claim that we understand the behaviour fully, we have to know how these basic units are put together—not just when a muscle contracts, but why the hen flutters at one moment and runs sideways the next. We have to know how the 'low level' programs of instructions to muscles are put together into 'higher level' programs that take account of the goal-directedness of the behaviour.

Much of this chapter will be concerned with what we know about the rules of animal nervous systems. We will run the whole gamut from low-level to high-level programs, from the circuit diagrams of physiology to behaviour level programs which are often best studied by a 'whole animal' approach, an approach which has the virtue of not damaging the animals which are being studied. We will pay particular attention to ideas and problems which are common to physiologists and ethologists, so as to make it easier to link together the various ways of studying behaviour. The first such problem which concerns everybody is that of the initial description of the behaviour itself and the choice of the most useful unit to pick out and ask questions about.

3.2   Units

3.2.1   *The Fixed Action Pattern as the unit of behaviour*

One of the most influential attempts to define a unit of behaviour was that of Konrad Lorenz and Niko Tinbergen. Lorenz (1932, 1937) argued that each species has a repertoire of 'Fixed Action Patterns' which are just as much a diagnostic feature of that species as morphological characters like crests or hooves or coat colour. The courtship patterns of ducks and the nest-building movements of hummingbirds were some of the examples he cited. The crucial feature of a Fixed Action Pattern, however, is that Lorenz did not define it simply by whether or not an action appeared to be 'fixed' or stereotyped, that is, repeated in the same way on every occasion (a description of the output itself). He went much further and proposed a 'package deal' definition which included, as well as stereotypy, a hypothesis about a particular kind of motor control (program) and also a particular mode of development (how the animal's nervous system acquired the program in the first place).

We will worry about the problem of how stereotyped a behaviour pattern has to be before it qualifies on these grounds as a Fixed Action Pattern in the next section. What we are going to do here is to look at Lorenz's ideas about the control of a sequence of behaviour, for it was this that became one of the most important characteristics of the Fixed Action Pattern or FAP (Hinde 1970). Lorenz hypothesised that whole sequences of instructions to the muscles could arise from within the nervous system itself. An FAP might be triggered by the environment but, once it was set going, it could carry on through to the end without any further input from the environment, like a clockwork toy going through its tricks one after the other. He thus distinguished it from a chain of reflexes in which performance of the first small step in a movement would stimulate the sense organs that activate the muscles to produce the next step and so on. The patterning of an FAP, on the other hand, was held to come from within the nervous system without the need for external 'prodding' or stimulation by sense organs at each stage. The unit of behaviour could then be said to be that sequence which was the result of one of these central nervous motor 'programs'.

Lorenz and Tinbergen also added that, although the actual sequence making up the FAP was supposed (by definition) to remain impervious to environmental feedback while it was in progress, its orientation or taxis component did not have to be. The greylag goose (*Anser anser*) retrieves eggs that have rolled out of its nest by stretching its neck and pulling an egg towards it. The actual retrieving movements which the goose makes are always the same, regardless of whether the goose is trying to roll an egg or a wooden cube (if one is put there by an ethologist). These movements they called the Fixed Action Pattern. However, the orienting response—the lateral movements of the bill by which the goose prevents the objects from rolling off sideways as she is retrieving them—are affected by feedback from the object being rolled. Thus if the object does not roll off to one side, the goose very sensibly does not try to stop it doing so. In their view, then, the basic FAP, which was fixed, could have an orienting component which was modifiable.

At the time, the evidence that complex sequences of behaviour could derive from patterns arising within the nervous system was not very convincing. For one thing, the physiological basis for supposed FAPs such as egg-rolling was completely unknown. Some thirty years later, however, neurophysiological evidence, particularly from invertebrates, has begun to accumulate suggesting that some nervous systems are indeed capable of generating complex motor programs almost, if not entirely, 'off their own bat'.

One of the most striking examples of this is the escape swimming response of the sea slug *Tritonia*. After only the briefest touch by its main predator, a voracious starfish, *Tritonia* goes into a stereotyped series of swimming movements, flapping its body up and down in a way which, although clumsy, is usually enough to carry it away from the starfish. The response lasts for about 30 seconds and is brought about by the alternate flexing of the dorsal and ventral muscles of the body wall.

Now, before we can conclude that the 'escape swimming' instructions to these two groups of muscles are in fact originating within the animal's nervous system, we have to exclude an alternative possibility, namely that the muscles are being activated in a chain of reflexes, each one triggered by previous muscular movement. What this would mean would be that, say, an initial withdrawal movement to the stimulus of the starfish might activate stretch receptors in the muscles to send impulses to the brain that then direct the elongation of the body which in turn results in other stretch-receptor messages coming back to the brain to initiate contraction by other muscles and so on.

This possibility was excluded by removing the brain of a *Tritonia*, putting it in a dish by itself and recording the electrical activity in its large cells (many of them up to 1 mm in diameter). The 'escape swimming' program could be initiated in such an isolated brain by stimulating with a brief shock some of the peripheral nerve stumps still attached to it. The brain then responded in much the same way as it was known to do in the intact animal, giving a 'play-back' of the instructions it would be issuing to the dorsal and ventral muscles, only of course there were no muscles there to receive them (Willows 1971; Dorsett *et al.* 1973). Feedback from the muscles thus seems unnecessary for the brain to continue to issue its instructions. It may be that in the intact animal the brain does respond to things that are happening in the muscles or elsewhere in the body but the point is that it does not have to, for the nervous system is itself capable of generating a long series of instructions, much as Lorenz had suggested many years before.

Nor is this an isolated case. There are numerous examples from both invertebrates and vertebrates which illustrate the same thing. The rhythmic mouth movements of feeding water snails such as *Helisoma* are brought about by the ordered contraction of 25 pairs of muscles. The program which controls these muscles, that is, the series of commands as to which ones should contract when, originates within two knots of nerve cells, the buccal ganglia. As in the case of *Tritonia*, these ganglia can be completely removed from the rest of the animal and yet they can still go on issuing the same commands as they did in the intact animal (Kater & Rowell 1973). Some feedback from the muscles appears to be possible here since extra power can be applied to the mouth if the snail encounters a particularly hard bit of food, but it is not necessary for occurrence of the basic feeding rhythm. The same seems to apply to feeding in other molluscs such as the slug *Limax* (Gelperin *et al*. 1978; Reingold & Gelperin 1980) and indeed to *Tritonia* (Bulloch & Dorsett 1979). Other rhythmical movements, such as singing in male crickets, which consists of patterned pulses of sound brought about by one wing being rubbed over another, also seem to depend on sequences of muscle instructions originating inside the nervous system (Bentley 1977) as does a major part of courtship in the grasshopper *Gomphocerippus*, which is a very complicated affair involving singing, head shaking, swinging the antennae and gesticulating with the hind legs (Elsner 1973).

Even in vertebrates, movements can result from centrally generated patterns of muscle instructions. These include walking in tetrapods and swimming in fish and amphibians, although, as might be expected, there is also a considerable role for feedback and adjustment while the movement is in progress, as the animals are travelling through often unpredictable environments and have to be able to cope with obstacles and unexpected pitfalls (Stein 1978; Grillner 1975; Delcomyn 1980). The call of the leopard frog (*Rana pipiens*) can be traced to movements of the larynx controlled by instructions from the brainstem. Even where there is no peripheral feedback from the laryngeal muscles or any other source to the brainstem, it still goes on issuing the sequence of 'calling' instructions (Schmidt 1974). When mice groom themselves, the front paws are moved in characteristic ways over the face. Even when the mouse is not receiving any sensory input from its face, the grooming actions still occur in the usual sequence (Fentress 1976).

The nervous system is also capable of giving rise to even more complex sequences of behaviour, lasting minutes or even hours. The giant silkmoth (*Hyalophora cercropia*) emerges from its cocoon in three distinct phases. First there is a period of intense activity lasting about half an hour in which the abdomen is rotated and twitched violently. Then follows a period of rest, also lasting half an hour, and finally there is a second burst of activity as the moth finally emerges. These same three phases are shown by a de-afferented nerve cord on its own, once the response has been triggered off by the right hormone (Truman & Sokelove 1972; Truman 1978).

There is, then, a solid basis for Lorenz's claim that the patterns for many behavioural sequences arise from within the nervous system, without the need for constant feedback from the environment. Does this mean that we can now go ahead and use the FAP as the basic unit of all behaviour? There are unfortunately several quite powerful reasons why we cannot (Barlow 1968, 1977). The first and most obvious is that, although we know the physiological basis of *some* behaviour patterns and can state with confidence that *some* aspects of them are independent of environmental or sensory feedback control, this is not true in all, or even the majority, of cases. Ethologists have tended, on the whole, to work with birds, fishes and mammals, while the greatest advances in 'circuit breaking' have come from the study of invertebrate nervous systems. This means that there is a great deal of behaviour, such as courtship in fish and birds, for which we have little knowledge of the physiology. Yet we want to study such behaviour and ask questions about it. We therefore need some way of defining behaviour patterns which does not depend on internal mechanisms.

A second important reason why the FAP will not do as the unit of all behaviour is that even in those cases where there is considerable endogenous nervous activity, there usually turn out to be quite large effects of environmental feedback as well (Bentley & Konishi 1978). We have seen how, in the feeding of *Helisoma*, the centrally generated program can be modified by external circumstances and that the same is true of many other action patterns such as swimming (Grillner & Wallen 1977). The result is that it is often difficult to make a clear-cut distinction between centrally generated behaviour patterns which have some environmental input on the one hand, and a chain of reflexes on the other (Hoyle 1976). Most animal behaviour simply does not fall neatly into packages described as the output from one centrally generated program, and many so-called Fixed Action Patterns turn out on closer analysis not to be so fixed after all (Moltz 1965; Barlow 1968;

Stamps & Barlow 1973) but to vary considerably from occasion to occasion. Even though Lorenz did a great service by stressing the creative activity of the nervous system at a time when the prevailing orthodoxy was to view all behaviour in terms of reflexes, the control of behaviour sequences has turned out to be far more complex, and the sequences themselves far more variable from case to case, than can be accommodated by the FAP as the universal unit of behaviour.

Yet another objection to the use of the term FAP concerns the postulated development of FAPs (Moltz 1965). Part of Lorenz's original definition was that it should be 'innate' or 'genetically encoded in the nervous system'. Indeed the German word for FAP, 'Erbkoordination', means literally 'hereditary coordination'. The controversy which surrounded the word 'innate' (Bateson 1983) resulted, for a time, in ethologists being rather reluctant to talk about innate behaviour at all and, by association, the FAP suffered a somewhat similar fate.

For all these reasons, then, the term Fixed Action Pattern is not very widely used today by ethologists (although, perhaps paradoxically, it is now much more commonly used by neurophysiologists). George Barlow's (1968, 1977) sensible suggestion that we should instead use the term Modal Action Pattern or MAP (to stress the average or modal nature of much behaviour and to get away from the idea of absolute fixity) has, for some reason, not been widely accepted. Most people talk about 'action patterns', 'behaviour patterns' or simply 'acts' to avoid any assumptions about underlying control or implications of innateness. But that, of course, simply brings us back to the problem we encountered at the beginning. What are 'behaviour patterns' or 'acts'? If we cannot use the FAP, what *is* the unit of behaviour?

3.2.2   The statistical definition of a unit of behaviour

The most widely used definition of a unit of behaviour is, either implicitly or explicitly, a statistical one. The key to understanding what is meant by defining a behaviour pattern statistically lies in the word 'pattern' itself. Pattern means some kind of non-randomness, regularity or predictability. If we were asked to continue the sequence 1 2 3 4 3 2 1 2 3 4 3 . . . we would have no difficulty in doing so. We could *predict* the next number in the series from having detected a *pattern* in the previous ones. There is thus a very close connection between pattern and predictability and indeed what we mean by a behaviour pattern is precisely that it follows a predictable sequence. A chick walks by moving its legs and head, and film analysis (Bangert 1960) can reveal the exact details of these movements. We do not, however, go around saying that the chick 'forms an angle of x between its body and its right leg, and angle y between body and left leg, the head–tail length is z at time t then, at $t+1$ s, the angle between the body and the right leg is x' . . .'. We say simply that the chick is 'walking'. Now the reason that this is valid (as well as being a great deal shorter) is that there is no *need* to describe leg movements and head movements separately or to see leg positions at one moment as totally independent of those a few seconds later. They follow predictable patterns. The head is jerked rapidly forwards in co-ordination with the legs. If the left leg has just begun to move forwards, it will continue to do so for a while and then the right one will be moved and so on. If we know what one part of the animal is doing, we can predict with a fair degree of accuracy what other parts will be doing and also what will be happening at various times in the future. 'Walking' is sufficiently regular that we can recognise it as a coherent whole, like a repeated motif on a wallpaper.

It happens that in practice we generally recognise patterns of behaviour such as walking intuitively: our 'computer in the head' is a very good statistical pattern recogniser. But it is possible to make this pattern recognition process rather more objective and external than this. Chicks drink by dipping their beak into water and then raising the beak. A videotape slowed down shows that the chick's eye traces a characteristic path as the bird drinks (Fig. 3.1a). If several of these traces are superimposed, it is clear that there is a considerable degree of regularity. It is not quite the case however that once you've seen one drink you've seen them all. The way a chick raises its beak to let the water trickle down its throat is very similar from one drink to the next, but the way it lowers its head to the water at the beginning is very much more variable, since the bird may pause with its head part of the way down to the water in a somewhat unpredictable way.

Another way of putting this would be to say that if we were watching chick drinking, there would be some times, such as when the bird was lifting its beak out of the water, when we could predict very accurately what was going to happen in the next second or so: the beak would almost certainly continue to gain height. There would be other times, however, such as when the bird was lowering its beak into the water at the beginning of a drink, when we would be very hard put to it to say whether it was going to continue to lower its head, to pause awhile, or to raise its

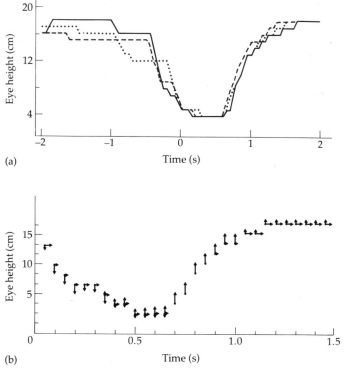

(a)

(b)

Fig. 3.1. (a) Superimposed graphs of the height above the ground of a chick's eye during three successive drinks, lined up on the moment when the bird's bill strikes the water (time 0). (From Dawkins & Dawkins 1973.) (b) Diagrammatic representation of changes in a computer's predictions over time during one particular drink. The direction in which the arrows point represents the direction in which it is predicted that the eye will move during the next six 'frames' of the videotape. The length of the arrows represents the probability associated with these predictions based on the computer's previous information about how chicks drink. For example, for most of the frames of the upstroke, the probability is 1.0 that the average eye height in the next six frames will be greater—this is a highly patterned, predictable part of the drink.

head again and abort the drinking action altogether. Figure 3.1b shows such predictions made by a computer 'watching' a chick drinking (actually it was given a series of numbers representing the height of the chick's eye above the ground). Some parts of the drink are clearly more stereotyped or predictable than others. So 'drinking' is not an absolutely fixed and regular sequence. But it does possess a degree of predictability, with some elements such as the 'upstroke' being particularly resistant to interference and remaining much the same even if the chick is drinking something it finds distasteful (Dawkins & Dawkins 1974). If we compare drinking in chicks which have been deprived of water and chicks which have not, we find that the main difference is in the intervals *between* drinks, the drinks themselves being relatively unaffected, again suggesting that 'the drink' does form a kind of unit, even if not a perfect or indissoluble one.

Other intuitively recognised behaviour patterns, too, have shown a high degree of stereotypy when analysed statistically. The 'strut' display of the male sage grouse, in which the bird shows coordinated movements of its wings and oesophageal sac and produces a strange sound at the same time, is very constant in duration, whereas the interval *between* struts is much more variable (Wiley 1973). One of the courtship patterns of the goldeneye duck (*Bucephala clangula*), the head-throw display, is also very fixed in duration. It lasts 1.29 s, with a standard deviation of only 0.08 s (Dane *et al.* 1959).

It might be said, however, that such statistical analyses of behaviour patterns using slowed-down film or videotape really tell us nothing that we could not have gathered intuitively anyway, but it is nevertheless instructive to externalise the process of recognising behavioural units, if only because it emphasises the similarity of the processes of looking for pattern on very fine time scales of fractions of seconds and on much longer ones of hours or days. It is probably no accident that most of the behaviour patterns that have been described last between 0.1 and 10 seconds, with a mode of 1 s, because this is a particularly easy time scale for the human brain to pick up (Schleidt 1974). We are much less sensitive to patterns with a longer repeat time, say one or two hours, and may *have* to resort to statistical analyses to become aware of them at all.

If we do externalise our recognition of behaviour patterns,

there is also less likely to be disagreement between two observers as to what an animal is actually doing. It is quite possible for one person to watch a chimpanzee for an hour and note down 40 different behaviour patterns and for another to watch the same animal for the same time and end up with only 30 patterns on his list. The second observer might have lumped together actions that the first observer saw as distinct. If they could both agree on clear rules or algorithms (ideally those that a computer could use) to define each action, they would very much reduce their chances of disagreeing, and if they used slowed-down film, they might see nuances of gesture that both would have missed with the naked eye.

In practice, most studies of behaviour bypass the formal statistical analysis of behaviour units and rely on the 'computer in the head' to recognise intuitively the basic components of the animal's repertoire or 'ethogram'—acts like pecking, scratching and so on. The pattern is sought among these basic units, using rather more overt statistical methods. The search for pattern, however, is going on at both these levels, whether we are consciously aware of it or not.

3.3   Sequences

3.3.1   Sequences of action patterns

Given that we have decided, intuitively or otherwise, what our units of behaviour are, we then come on to the question of how behaviour patterns are linked into sequences. Why does the animal now perform this behaviour pattern, now that? Physiologists are discovering a great deal about how this occurs. Carlson and Bentley (1977), for instance, investigated what they called the neural orchestration of moulting in crickets. A cricket takes about four hours to shed its skin and during that time almost every muscle of its body is brought into play in a sophisticated series of operations. During this sequence, 48 low-level motor programs are activated, some in series, some simultaneously. This was discovered by recording the activity of individual muscles and the nerves leading to them. Powerful though these techniques are, however, they are not the only ones we can use. We do not have to open an animal up to find out how its body works. We can use a 'whole animal' approach,working out what is going on inside the animal from observations from the outside without piercing its skin at all. What this approach involves and what we can learn from it we will illustrate with a study of a talking elephant.

3.3.2 Horton—a case history

This section is not a joke. The talking elephant—a toy that says one of several sentences when a string is pulled and released—is used in a perfectly serious way to teach students the principles of using observations to deduce the organisation of behaviour (Hailman & Sustare 1973). The mechanism inside the toy is, of course, quite different from that of, say, a singing bird, but understanding how we can use observations of its behaviour to deduce something about how it works clears the way for understanding the same principle for more complex behaviour. Horton was one such talking elephant. Each time his string was pulled, he uttered one of ten homilies such as 'An elephant's faithful, one hundred per cent' or 'A person's a person, no matter how small'. These utterances, incidentally, are perfect examples of FAPs by all definitions: they were always the same, and once started they always went to completion without any environmental interference. They were hard-wired into the animal.

A class of students at the University of Wisconsin began by pulling Horton's string 10000 times and noting down what he said. From this they could see that he was about equally likely to say each sentence in his repertoire, but that having opted for one sentence type he then became more likely to say certain other ones. Thus he was very unlikely to repeat himself (A followed by A) but rather likely to follow A with B or I. From these observations, the students (with a little help) hypothesised that Horton's mechanism consisted of a pointer moving around a circle; wherever the pointer stopped, Horton would utter that vocalisation. Their model of the sequence round the circle was derived by writing down all the sentences on a piece of paper and then drawing arrows from each one to the three other sentences that followed it with lowest probability. Thus A was most unlikely to be followed by itself or H or G. The resultant spaghetti was then tidied up to form the simplest circular shape possible (Fig. 3.2). The interesting thing was that this model, having been derived on the

basis of the *least* frequent transitions from one utterance to the
next, turned out to be an excellent predictor of the *most* frequent
ones as well. Thus A and H had been placed near each other on the
circle on the grounds that Horton almost never said these two
consecutively. But if Horton had just said E, his next sentence was
most likely to be either A or H. Two separate sorts of evidence—
the low probability of transition from one to the next and the high
probability of being chosen to follow the same previous utterance
—thus both gave rise to the same arrangement of sentences
around the circle, a good indication that the model was at least
internally consistent.

Many other things were deduced about Horton—the direction
in which his pointer moved, for example, and how far it moved on
each string pull—all from observations of the order in which he
said things. On the basis of the information they had collected, the
class speculated about actual physical mechanisms. The two most
likely seemed to be a conventional recording disc with grooves
spiralling into the centre, and a cylinder with spiral response
grooves. Pulling the string in either case would wind a mechanism
that rotated the cylinder or disc until a phonograph needle (the

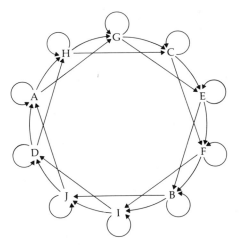

Fig. 3.2. Diagram showing the connectivity of ordered pairs of Horton's utterances
where there is a low probability of one following another (e.g. A is unlikely to be
followed by either H or G). Recurrent arrows indicate that no response is likely to
follow itself. The circle suggested a model of how Horton's mechanism worked.
(From Hailman & Sustare 1973.)

pointer) was over a particular groove (one for each utterance). Releasing the string resulted in the needle being dragged along the groove as the cylinder or disc rotated back again.

At the end of the study, an autopsy was performed on Horton to reveal a recording disc with spiral grooves. This result was received with 'a hearty cheer' from the student onlookers, although the discovery of the actual physical mechanism could be said to have added little to the knowledge of behavioural organisation they had already gained. The important thing was to have discovered what the mechanism did, not what it was made of.

Notice, though, that when we say that the students had discovered 'what the mechanism did' we do not mean that they could predict exactly what Horton would say on the next pull of his string. They could give only a probabilistic prediction. Their understanding was thus not 'deterministic' but 'stochastic' (which means the same as probabilistic). The reason why studying Horton was such a good preparation for studying real animals is that their behaviour, too, often seems to consist of probabilistic, not deterministic, sequences (Metz 1974; Cane 1978). For example, if a turkey cock has just 'gobbled', we cannot say exactly what it will do next, but there is a very high chance that it will 'strut' in the next few seconds (Schleidt 1964). Or, if a male zebra finch has just preened itself, the next thing it does will almost certainly be another preening movement, but it is not possible to say whether it will be ruffling the feathers or scratching (Slater & Ollason 1972). Frequently, we have to use statistical tests to be sure that there really is an increased probability of a behaviour pattern occurring when another has just been performed (for explanations of what these tests are see Slater 1973; Fagen & Young 1978). Horton, with his relatively simple behaviour, provided a good opportunity for the Wisconsin students to become familiar with these methods before they went on, with great enthusiasm, to apply them to the much richer behavioural repertoire of male mallard ducks (Hailman & Sustare 1973).

3.3.3   From toy elephants to real animals

Horton was used as a teaching aid to illustrate how much can be learned from systematic and quantitative observations of what goes on outside the body, albeit one which is much simpler than any animal. But the differences between Horton and real animals are also very instructive and help us to understand some important concepts. For one thing, most real animals are highly dependent on *time*. It made no difference to Horton whether he was left for 10 seconds, 10 hours or 10 days between string pulls. The time interval did not affect *what* he said next. Most animals are quite the opposite. In the small South American fish *Corynopoma*, for instance, the male performs a series of courtship actions such as Chase, Twitch and Quiver, while the female looks on fairly passively. If the male has just done a Quiver and goes on to the next action within five seconds, it is possible to predict fairly accurately what that action will be. But if more than five seconds elapses, it becomes impossible to distinguish the sequence of the fish's behaviour from a random sequence made up of all his courtship actions. The pattern has disappeared and, literally, paled into insignificance with the passage of time (Nelson 1964). Similarly, when mistle thrushes (*Turdus viscivorus*) sing, the notes of their songs are highly patterned—certain notes are particularly likely to follow certain other ones, as in music. The longer the bird pauses between notes, however, the more this patterning begins to decay (Isaac & Marler 1963).

Sometimes, indeed, the timing of one action in relation to another is itself a valuable first hint as to the underlying mechanism. In *Aplysia* (another sea slug), for example, swimming consists of alternate opening and closing of parapodia (flaps) on each side of the body. Slowed-down video film of freely swimming *Aplysia* showed that a wave of movement starts in the front parapodia and moves towards the back end of the body. At a given temperature, the period of this is fixed (that is, there is a constant interval between the beginning of one wave and the beginning of the next). But as the temperature is lowered, the wave is seen to disintegrate into two components: the period of the oscillation itself and the phase relation between the movements of the front parapodia and those further back. As a result, at low temperatures, both the front and back parapodia go on moving, but the phase relations between them change. The fact that temperature has a different effect on these two components of swimming suggests that they have different underlying mechanisms (van der Porten *et al.* 1980). As a water snail feeds, it not only moves its mouth, but also swings its head from side to side to produce a broad swathe through the alga on which it is feeding. Changes in the direction of the head swing, however, occur at very set times in relation to the mouth movements—only when the mouth is closed and never during the time when the radula is actually scraping the alga (Dawkins 1974). The head is swung to and fro much more frequently when the snail is moving slowly than when it is moving faster, suggesting that the mechanisms controlling mouth movements, head movements and locomotion are all closely co-ordinated.

Unlike Horton, almost all animals behave differently at different times of the day—they show circadian or 'about a day' rhythms, which may persist even when the animals are kept in constant light and deprived of all obvious environmental cues that one day has ended and the next begun. This has led to the suggestion that animals have 'clocks' built into their nervous systems (Harker 1974; Saunders 1977). Animals that live in tidal zones may have additional clocks based on the moon (Palmer 1975), and yearly clocks may help insects to emerge at the right season (Saunders 1976) and migratory birds to set off on their journeys at the best time (Gwinner 1972; Berthold 1974). Time, then, is an extremely important factor for animals, one which we have to take into account when we observe their behaviour, when we design experiments and when we think of the underlying mechanisms.

A second very important difference between Horton and real animals lies in something we have touched upon briefly at the beginning of this chapter—namely the fact that much animal behaviour appears not as a rigid sequence, but directed towards some end-point or goal.

3.4   Hierarchies

3.4.1   Goal-directed behaviour

We come back again to a crucial point about behaviour: animals behave *appropriately* in a world which is largely made up of unpredictability and novelty. Not all of this world is unpredictable, of course. Some aspects of it are sufficiently regular and constant that appropriate behaviour can be achieved by relying on set sequences which do not allow the environment to feed back its effects on the animal and alter the course of behaviour. We have already met this idea in the form of the Fixed Action Pattern. Even on this level, on time scales of a few seconds, feedback and adjustment are usually able to modify the basic patterns of feeding or locomotion precisely because the unexpected may happen even when the time horizon is only a few seconds ahead. How much more necessary they are when this horizon is extended to minutes or hours, when the animal is, say, tracking a moving target (Chapter 2) or fighting an opponent of unknown strength.

We often find, then, that the later parts of a sequence are influenced by the results of the earlier parts, perhaps that the sequence is actually brought to an end by its own consequences. Some examples will illustrate some of the ways in which this can happen.

The great golden digger wasp (*Sphex ichneumoneus*) digs burrows into which it puts insects as food for its young. The depth of a burrow is important: if it is too short, the wasp larva will be caught in topsoil which becomes so hot in summer that it would almost certainly be lethal; if it is too long, the larva may not develop properly from being too cold and the adult would in any case have wasted time on digging that it might have spent getting on with the next burrow. The wasp achieves the 'right' burrow depth not by digging for a fixed amount of time or making a fixed number of digging movements but by responding to the actual burrow itself. If her burrow is artificially lengthened, the wasp takes this into account and digs less herself (Brockmann 1980).

Spiders building their webs also seem to rely heavily on feedback and achieving certain states of the web before going on to the next stage. Orb-web spiders build a frame with spokes radiating out from the centre and then lay a spiral line over the spokes. Removing some of the radial spokes from the web of the cross spider (*Araneus diadematus*) while the spider is building it does not affect the spider's ability to complete a web of the normal sort—it simply repairs the damage before continuing with the next stage (Peters 1970).

Even more complex sequences of building are shown by weaverbirds, which make an intricate basketwork nest often of elaborate shapes which help to prevent snakes and other predators getting in. Some male weaverbirds are polygamous and may have several nests under construction all at once, so that any simple 'fixed sequence' of building can be ruled out straight away. One species, the Baya weaver (*Ploceus philippinus*) builds its nest in the sequence shown in Fig. 3.3 and is able to make extensive repairs to mutilated nests, often exactly replicating the original structure (Crook 1964).

It is not only the inanimate physical environment which can provide feedback. Early parts of a sequence may affect the behaviour of another animal, which in turn alters that of the first. The male smooth newt (*Triturus vulgaris*), for example, does not deposit his spermatophore until late on in the courtship sequence when the female has shown her readiness to pick it up by touching his tail with her nose, a condition which can be easily mimicked with a model female since a mechanical touch seems to be what is important (Halliday 1975). The courtship displays of the male

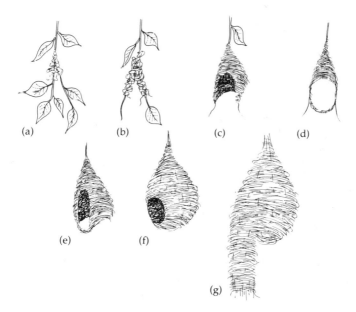

Fig. 3.3. Stages in the building of a nest by the weaverbird *Ploceus philippinus*. (a) Wad of loose stiches knotted to support; (b) wad extended to form top of ring; (c) wad in the form of a cone with horns; (d) completed initial ring; (e) fresh 'helmet' stage; (f) completed egg chamber, no tube as yet added; (g) finished nest. (From Crook 1964.)

normally induce this behaviour in the female, which in turn stimulates the male to deposit the spermatophore.

To show that behaviour is affected by its consequences is one thing. To claim that it is therefore 'goal-directed' is quite another. Or is it? Do these examples show goal-directed behaviour just because there is evidence of feedback? Thorpe (1963) argues that goal-directedness or purpose means a striving after 'a future goal retained as some kind of image or idea'. Since we can have no direct knowledge of whether an animal has any ideas or images at all, the most we can hope to do is to describe it 'as if' it did, rather in the way that we can describe a Cruise missile behaving 'as if' goal-directed when homing in on a target. Thorpe then argues that one of the most important criteria for establishing whether animals are behaving as if they have goals is whether they show 'variable means to an invariable end', as did our hypothetical chicken mentioned earlier, trying out various strategies to achieve access to the food.

Hinde and Stevenson (1970) warn against the dangers of being too trigger-happy with the word 'goal' and the diversity of mechanisms the word may be used to cover. In the case of the weaverbirds, for instance, the ability to repair damaged nests might suggest that the birds have an image of what a completed nest should look like. Collias and Collias (1962), however, pointed out that the actual movements that the bird makes in both construction and repair, such as winding strips around twigs or poking them into the nest mass, are very stereotyped. The ability to repair a nest seems to depend on the bird being able to orient itself properly within the nest. If it stands in the right place and carries out its stereotyped building movements, the result is a repaired nest, but there is no necessary implication that the bird is in any sense comparing the nest's present structure with some kind of ideal pattern. In other words, the 'rule of thumb' which the animal is following could be very simple indeed and yet achieve a complex end state or goal.

This is not to say that rules of thumb are always simple. Rats placed in a maze with eight arms radiating out from the centre can learn to visit each arm just once to find food, somehow avoiding going down the same arm twice. Their rule of thumb is not the obvious one of avoiding arms in which they have left a scent trail or going round visiting each arm in turn. To explain the sequence in which they visit the arms, it seems necessary to postulate that they have an internal spatial map or image of the maze (Olton & Samuelson 1976).

Somewhat similar feats are accomplished by marsh tits (*Parus palustris*), which store hundreds of seeds in their territories and then later find them again (Cowie *et al.* 1981; Sherry *et al.* 1981). The seeds are retrieved in a sequence which is neither triggered by visual or olfactory cues nor a systematic search, but again seems to involve an internal representation in which the bird 'ticks off' each seed as it finds it. We have here left the realm of simple stimulus–response and approached something more like Thorpe's images and ideas.

We turn, finally, to another very important concept that spans the physiological and 'whole animal' views of behaviour, one which has in fact hovered in the background all through this chapter and which we must now look at rather more closely.

3.4.2 The hierarchical organisation of behaviour

A 'hierarchy' means simply that there are some elements in a system that are higher than or superior to other elements. To some people, it is necessarily true that animal behaviour is hierarchically arranged. To others, it is false, or at least likely to be false. The reason for this disagreement arises from a confusion between two quite different usages of the term hierarchy. On the one hand, there are hierarchies of *embedment* (Nelson 1973) or classification, in which to say that one element is superior to others means that it includes them. Thus the Class Mammalia is superior to (because it includes) pandas, wolves, etc., and 'reproductive behaviour' in a bird could include building a nest, courting, etc. Using this kind of hierarchy, the choice of whether to describe behaviour as muscular movements or as 'reproduction' would be largely a question of how much detail is required.

On the other hand, there are hierarchies of *connection* or control. Here, to say that an element is superior to others means that it controls them or gives orders to them, as in the army. For example, there are giant nerve fibres in the crayfish which, when stimulated electrically, trigger a sudden bending of the abdomen that propels the animal backwards and which have therefore been labelled 'command cells'. In *Tritonia*, the escape swimming response which we have already discussed appears to be initiated by a pair of large nerve cells, known as C2s, which act as 'neural push buttons' and trigger the escape program (Getting 1977). These, too, have been called command cells to indicate that they do not themselves generate the patterns of command to the muscles —they simply trigger the activation of other cells which do (Kennedy & Davis 1977), rather in the way that a colonel could issue the command to retreat and leave the details of how to do this to officers of lower rank.

Although the term 'command cell' is usually attributed to Wiersma and Ikeda (1964), Hoyle (1978) points out that the idea behind it is similar to the hierarchical scheme of behaviour proposed to Tinbengen (1951), in which motivational energy was held to flow from higher to lower centres (Fig. 3.4). We can accept the criticisms of motivational energy (Chapter 4) and still retain the idea of a hierarchy of causal influences as a valuable one in the explanation of behaviour. For example, there might be a hormone which triggers off several different actions and in this sense could be said to be 'in command' of them. Alternatively, there may be neural hierarchies—cells or groups of cells which are superior to others in that they control their times of activation. But the claim that behaviour is controlled hierarchically is only a hypothesis which has to be tested and if necessary rejected, unlike the hierarchy of classification we discussed earlier, which did not make any assumptions about underlying mechanisms. Herein lies the confusion. Just because it is possible to *describe* behaviour hierarchically does not mean that it is *organised* hierarchically on a physiological basis (Dawkins 1976). The control of behaviour may, for example, turn out to be complicated by feedback loops from 'lower' to 'higher' centres, clouding the clear-cut ideas of a one-way hierarchy of command. Nevertheless, there are some observations of behaviour which do strongly suggest the operation of hierarchical rules within the nervous system.

We have stressed repeatedly that most behaviour is not simple fixed sequences of movement. One reason for this is that animals often behave as if directed towards some goal or end-point, in the sense that they employ variable means to a constant and identifiable end. These variable means (such as flying, running or walking to get at food) could therefore be seen as substitutes for one another in that the animal does one *or* the other and still achieves the goal. Goal-directedness thus implies that there are sets of substitute or mutually replaceable elements in the animal's behaviour. Even when we are unable to identify a specific goal, it often turns out that behaviour does seem to be made up of these mutually replaceable sets. For example, when crickets (*Teleogryllus*) groom themselves, the order in which they perform their

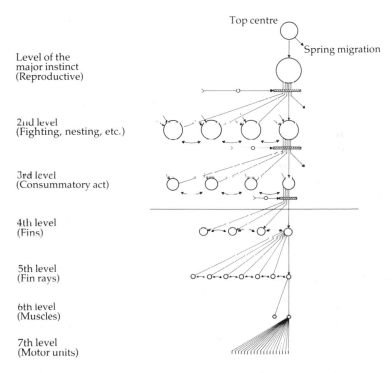

Level of the
major instinct
(Reproductive)

2nd level
(Fighting, nesting, etc.)

3rd level
(Consummatory act)

4th level
(Fins)

5th level
(Fin rays)

6th level
(Muscles)

7th level
(Motor units)

Top centre

Spring migration

Fig. 3.4. Tinbergen's hierarchical system as illustrated by the example of reproductive behaviour in the male three-spined stickleback (*Gasterosteus aculeatus*). At the top is the 'migration' centre which is influenced by hormones and changing environmental conditions and which causes the fish to migrate into shallow water. Migration occurs until the fish encounters a suitable territory. Here, stimuli such as warm, shallow water and suitable vegetation act to involve the next centre in the hierarchy, the 'reproductive' centre. Motivational energy then flows through this centre but cannot be passed to the centres next down the hierarchy (fighting, nesting etc.) until the fish encounters the right external stimuli (such as a rival male). When it does so, the motivational energy is channelled through these centres which, in the course of carrying out the behavioural acts, involve centres even lower down, right to the level of motor units. (From Tinbergen 1951.)

various grooming movements is, as is so often the case, a probabilistic sequence but not a simple linear one. If the cricket has just groomed, say, one of its forelimbs, it is most likely to groom next either an antenna or some other part of the *front* end of its body. On the other hand, if it has just groomed part of the back end of its body (such as rubbing its two back legs together), its next action will tend to be another back-end grooming movement. The grooming actions thus form two distinct clusters defined by chances of one act following another. There seems to be one set of rules controlling which cluster is performed and another controlling which action *within* a cluster is performed (Lefebvre 1981), in other words, a hierarchy of rules. Over and above the grooming rules is an even higher level of rules for switching between grooming and locomotion. The similarity between such a hierarchy of rules and the grammatical structure of human language has struck a number of authors (e.g. Marshall, quoted in Vowles 1970; Rodgers & Roseburgh 1979).

In conclusion, we can say that because animal behaviour is infinitely more complex than a series of fixed sequences, the programs governing it must in turn be more than simply instructions to the muscles of the sort 'do this, now that'. The programs have to take account of the goal-directedness of behaviour, of the fact that an animal may repeat an action a variable number of times to get at that goal or may switch and try something new altogether. It may have sub-goals to be completed before higher level ones can be achieved (Miller *et al.* 1960). Human computer programmers, trying to make their machines perform complex tasks, almost always employ hierarchically organised instructions for speed and efficiency. Perhaps not surprisingly, animal nervous systems, faced with even more complex tasks, seem to do the same thing.

3.5   Selected reading

The book by Oatley (1978) is worthwhile reading for those interested in the relationship between 'whole animal' and physiological approaches to the study of mechanisms of behaviour. Although mostly concerned with psychology, his message is very relevant to ethology too. Bentley and Konishi (1978) give a useful review of what has been achieved by recent advances in neuro-physiological studies of behaviour. Barlow (1977) is worth reading for a discussion of units of behaviour and where (and whether) Fixed Action Patterns fit in.

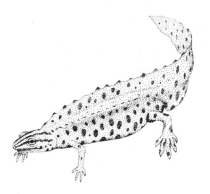

CHAPTER 4
MOTIVATION

TIM HALLIDAY

4.1   Introduction

When we say that an animal, or a person, is motivated to do something, we generally imply that their behaviour is driven or directed by some internal force or urge. This takes us into difficult ground, and ground that is very different from that considered in the first three chapters of this book. Chapters 1 and 2 dealt with the relationship between an animal's behaviour and the many kinds of external stimuli involved in its expression. Chapter 3 looked at how behaviour is organised in time. With this chapter, we start to delve into the animal's internal mechanisms and consider their role in determining behaviour. The very fact that the processes with which we shall be concerned are internal raises major difficulties, simply because they cannot be observed directly. They can only be inferred from overt behaviour and, in some cases, from a very limited range of physiological measurements and manipulations. This chapter begins with a discussion of what motivation is, how we set about studying it in an orderly, empirical fashion, and some conceptual problems that arise from studies of motivational processes. The remainder of the chapter examines a variety of animal activities and considers the extent to which we understand the internal processes that underlie them.

4.2   The study of motivation

4.2.1   What is motivation?

Much of the causal analysis of animal behaviour has been concerned with identifying the relationships between responses and

100

external stimuli. Many activities only occur in the presence of relevant stimuli; an animal cannot eat in the absence of food, for example. It can, however, look for food when it is not immediately available, and such food searching is behaviour that can only be explained in terms of internal processes. A more general point is that in many of their activities, animals do not always respond in the same way to external stimuli. Whereas a rat may withdraw its foot every time it steps on a hot surface, it does not eat whenever it is presented with food. Clearly, some internal process influences whether it responds to food. Such processes, whatever they may be, are the stuff of motivation.

If asked to provide a causal explanation of why an animal does not eat when presented with food, our most likely answer is that the animal is not hungry. This appears to provide an answer by invoking an internal process but in fact it really only describes what we have seen in different words and serves only to raise more questions. What is hunger and what are the physiological events that produce it? The essence of much research into motivation is the attempt to express concepts such as hunger in terms either of observable physiological processes or of rules which may help us to predict when a particular behaviour pattern will occur, such as those described in Chapter 5.

Before going further, we should consider briefly other possible reasons we might have used to explain an animal's failure to eat. One possibility is that the animal was engaged in another, perhaps more important activity, such as seeking a mate. The motivation of one activity is dependent on that of others, a major consideration which is discussed at length in Chapter 5. Another factor that may affect the way an animal responds to external stimuli is its stage of development. Young mammals being fed on their mother's milk do not respond to food they will eat as adults and, likewise, fully formed sexual responses generally do not appear until an animal is mature. Ontogenetic changes of this kind are not regarded as manifestations of motivational processes because they are irreversible; mature animals do not revert to immaturity. By contrast, being hungry, thirsty or sexually aroused are states that animals can experience frequently. Like maturational changes, learned responses to stimuli are usually also considered to be distinct from motivational processes because of their typically long-lasting influence on behaviour. Particular noxious flavours may be experienced once and avoided thereafter, and an animal's mating preferences may be restricted by early experience of its parents or siblings. Such effects cannot be ignored, however, but must always be taken into account in studies of motivation, as we will see in the case of feeding (section 4.6.1); as with other aspects of behaviour, motivation is subject to maturation and is influenced by learning.

The word motivation does not refer to a specific set of readily identified processes. Rather, it encompasses a broad category of phenomena that have in common that they are dependent on mechanisms internal to the animal and that they are reversible. Given the vagueness of the concept, we cannot assume that there is necessarily any similarity between the motivational processes that underlie different kinds of behaviour, or that similar processes operate in different kinds of animals. Essentially, an animal's motivational state is a 'black box' for which we may know a lot about the inputs (external stimuli) and the output (behaviour), but whose internal mechanisms are not observable. The task that faces those studying motivation is to reduce our ignorance about what goes on inside the box, whether it be the rules or the physiological processes that relate input and output, so that we can better understand their relationship to one another.

4.2.2   Intervening variables

Although hunger is a sensation of which we have direct experience, it is a hypothetical construct in the analysis of animal behaviour: it is something we cannot observe or measure directly, but can only infer from variations in overt behaviour. Since hunger is invoked to explain variations in response to food it is referred to as an intervening variable, because it has some modulating effect between a stimulus and its associated activity. Discussions of motivation are littered with intervening variables, such as thirst, sexual arousal, fear and, more generally, drives of one sort or another. The question we must ask if whether such intervening variables help in understanding motivation.

The basic problem with intervening variables is that any attempt to explain things in these terms involves a circular argument. We may infer that because an animal is eating voraciously, it has a high feeding drive. However, our only evidence for that assertion is the nature of the animal's behaviour, which is the very thing we are seeking to explain. It may appear that we break this circularity if we identify factors that influence feeding drive. For example, animals generally eat more readily if they have been deprived of food, and so we could say that food deprivation increases feeding drive which stimulates feeding. We can express this effect more economically, however, by simply saying that food deprivation stimulates feeding.

A hypothetical intervening variable may have some explanatory value if it enables us to simplify observed relationships between several stimulus inputs and a number of different behavioural outputs (Hinde 1982). Figure 4.1a shows three treatments that increase the tendency to drink in rats: water deprivation, feeding with dry food and injection with saline solution. The effect of all three treatments can be observed in three measures of drinking behaviour: the rate at which a rat presses a bar to obtain water, the volume of water it drinks, and the concentration of quinine (a

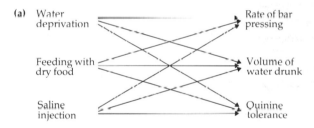

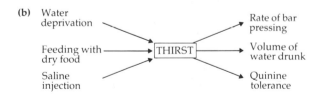

Fig. 4.1. The role of a hypothetical intervening variable in simplifying the relationships between causal factors and behavioural measures. (a) The three treatments on the left all influence the three scores of drinking behaviour on the right, in rats, giving nine causal relationships. (b) By introducing 'thirst', the number of causal relationships is reduced to six. (From Miller 1959, Hinde 1970.)

distasteful substance) it will tolerate in the water it drinks. The number of relationships between these treatments and measures can be reduced from nine to six by introducing thirst as an intervening variable (Fig. 4.1b).

Invoking a single variable, such as thirst, could therefore be useful provided that the motivational processes underlying the activity concerned constitute a single process. But there is no *a priori* basis for such an assumption and, indeed, there is abundant evidence that it is not justified in systems that have been studied in detail. The study of drinking in rats by Miller (1959) provides an example of such evidence. He gave rats saline solution by means of a tube inserted into their stomachs and then recorded the three measures of drinking mentioned above at various time intervals after the treatment (Fig. 4.2). One would assume that, as time passed, the rats' thirst would have increased and that, if their behaviour is indeed influenced by a single intervening variable, the three measures should be closely correlated. It is clear, how-

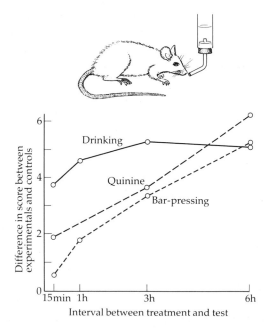

Fig. 4.2. Changes over time in three measures of drinking behaviour in rats following injection of saline solution into their stomachs. (From Miller 1956.)

ever, that they are not, as shown by the different shapes of the curves in Fig. 4.2. One measure, the concentration of quinine that rats will tolerate in their water, rises at a constant rate, as one would expect if thirst increased as a simple function of time elapsed since the treatment. By contrast, another measure, the volume of water drunk in 15 minutes ('Drinking' in Fig. 4.2), shows an increase only over the first three hours and then reaches an asymptote. On this score rats seem to be no thirstier after six hours than after three. One explanation of this effect might be that there is a limit to the volume of water a rat can drink in 15 minutes. These results suggest that different measures correlate rather poorly with one another, a finding which is contrary to the idea that these measures of drinking are all influenced through a single variable that we could call thirst.

In summary, when an intervening variable is invoked, we should not believe that anything has been explained; all that has been done is that some phenomenon that requires explanation has been identified. We cannot even assume that a unitary process is implicated. At best, intervening variables are labels for unknown physiological processes. The ultimate aim of much research into motivation is to identify and understand how such processes work, so that concepts such as hunger, thirst and drive become unnecessary.

4.2.3 Approaches to motivation

Motivation is a subject that has excited the interest of people from several different disciplines within the behavioural sciences. One major group are physiological psychologists who typically adopt a single-system approach, devoting their attention to the motivation of one activity, such as feeding, drinking or sex (see Chapter 6). They tend to work on a limited variety of animals, most often rats or pigeons, in highly controlled laboratory environments. Much of their work involves manipulating physiological variables as well as external conditions and recording the resulting behaviour. By contrast, most ethologists pay little or no attention to physiology but generally confine themselves to a more theoretical approach to the processes that occur between external stimuli and behaviour. Interested in a wide variety of animals living in complex natural environments, they are less concerned with what motivates any one kind of behaviour than with how different activities interact with one another (see Chapter 5).

Whichever of these approaches they represent, studies of motivation generally fall into one of three categories (Hinde 1982). The first involves analysis at a purely behavioural level; detailed observations are made of changes in behaviour and these are correlated with observed variations in external stimuli. Though this approach may lead to particular internal processes being postulated, little or no attempt is made to 'probe beneath the skin'. This essentially ethological approach has the obvious disadvantage that no precise information is obtained about the internal processes involved. Its main advantage is that, because work of this sort is usually conducted in natural or semi-natural conditions, it does not lose touch with the way animals normally behave in nature. The second category involves active manipulation of the animal's internal state, for example by brain lesions or hormone injections, so as to investigate its role in the control of behaviour. Because this physiological approach typically requires complex procedures, it usually has to be conducted in highly artificial circumstances. Thus, while it may be successful in pinpointing specific nerve centres or hormone levels as important factors in motivation, there is a danger that the animal's laboratory surroundings are so far removed from its natural environment that we can conclude rather little about how its behaviour is motivated in nature. The third approach is that involving modelling, and the advantages and problems of this kind of enterprise are discussed in Chapter 6. These three approaches should not be regarded as being mutually exclusive; ideally, each should complement the others and a full understanding of motivation may well involve all three.

4.2.4   Causal factors

In the more recent literature on motivation there has been a tendency to replace the word stimulus by the term causal factor. A causal factor is any event, process or change in some condition which can be shown to activate, sustain or inhibit a particular behaviour pattern. Causal factors may be external stimuli, such as the appearance of a predator, the presentation of food or the behaviour of a mate, or internal states, such as the level in the blood of a hormone or of glucose. Internal and external causal factors interact to elicit behaviour, a point graphically made by Lorenz (1950) in his hydraulic model of motivation (Fig. 4.3). He visualised the drive, or action specific energy as he called it, that motivates a particular activity building up, like water flowing into a tank, while an animal is not performing that activity. The behaviour is activated when the accumulated water is released, and this may occur in one of two ways. If there is a large accumulation

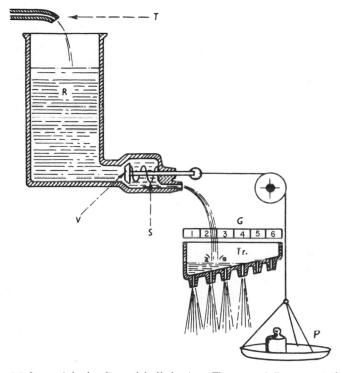

Fig. 4.3. Lorenz's hydraulic model of behaviour. The reservoir R represents the animal's drive level. Internal input as a result of deprivation enters through the tap T. No behaviour is expressed while the valve V is kept shut by the spring S. The valve is opened by a combination of the force exerted against S by weights in the pan P, representing external stimuli of varying strength, and by the weight of water in R. Different kinds of behavioural output are determined by a perforated trough Tr. If the combined force exerted by R and P is small, V will only partially open and only a small flow of water will leave Tr through output 1, representing the activity with the lowest threshold. Larger amounts of water will leave Tr through outputs 2 to 6, which represent activities with successively higher thresholds. (From Lorenz 1950.)

(shown as a big head of water in the model), it forces its way out through a valve. Alternatively, the valve may be opened by weights, the analogue of the appropriate external stimuli. Normally the two forces, internal and external, act in concert so that the valve will be opened by a combination of internal pressure and external force; if either is weak, the other must be correspondingly strong for the valve to open.

Lorenz's model gives a picture of behaviour being activated by both internal causal factors that provide 'push' and external causal factors that 'pull', such that its appearance depends on the combined strength of these push and pull forces at that moment. External causal factors thus vary in their effectiveness in eliciting behaviour, their capacity to do so depending on what is often referred to as their *cue strength* or, especially in the psychological literature, their *incentive value*. But whether an external causal factor with a particular cue strength elicits a response also depends on the strength of internal causal factors. Thus, food of rather low palatability may elicit feeding in animals that have been deprived of food for a long time, whereas animals that have recently eaten their fill require a particularly juicy titbit if they are to be tempted to eat again.

The important point that emerges from this way of looking at the causation of behaviour is that although a major aim of motivation research is to identify and understand the internal processes that influence behaviour, we must not ignore the role of external causal factors. In studying motivation, it is the *interaction* between internal and external causal factors that leads to behaviour and which must be our prime concern. This is perhaps the one valuable point that Lorenz's hydraulic model makes. In other respects, and especially in its vision of drive as some form of energy that accumulates and in its failure to include the effects of feedback from the consequences of performing behaviour, it was a very poor model of motivational systems (see Chapter 6).

4.3   Three studies of motivation

In this section, we look at three studies of motivation, chosen somewhat arbitrarily but which illustrate different approaches and raise a variety of issues. All three studies are concerned with sexual behaviour.

4.3.1   *Guppies*

Male guppies (*Poecilia reticulata*) court females with a variety of displays. Whether he displays at all and, if so, which display he uses, depends both on a male's reproductive state and on the size of the female presented to him. This interaction between internal and external causal factors has been analysed by Baerends *et al.* (1955). Rather conveniently, the colouration of a male provides an index of his reproductive condition (Fig. 4.4). Whatever his state, a male is more likely to display to a large female than to a small one. This is probably for the good functional reason that large females bear more eggs and mating with them therefore represents a greater potential increase in fitness to a male.

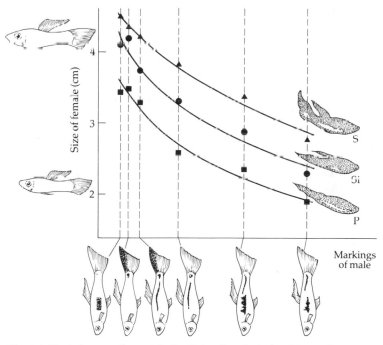

Fig. 4.4. The influence of external stimulation (female size) and internal state (indicated by male colouration) on the courtship behaviour of male guppies (*Poecilia reticulata*). The curves are for three different displays, posturing (P), sigmoid intention movements (Si), and complete sigmoid movements (S). Each point represents the size of female necessary to elicit a particular display. (From Baerends *et al.* 1955.)

The points in Fig. 4.4 represent the female size necessary to elicit three different male displays. As a male become increasingly sexually motivated, he responds to smaller females but, whatever his condition, larger females are necessary to elicit the full sigmoid display (S) than are required for intention sigmoids (Si) or posturing (P). What these results show is that males of low sexual motivation require a stronger external stimulus than those in which motivation is high, very much as Lorenz's hydraulic model suggested.

The curves linking the points in Fig. 4.4 are described as motivational isoclines by McFarland (1976). They link points of equal tendency to perform a particular display, the tendency being a function of both internal and external causal factors. In this example one cannot determine the exact form of the isoclines, because the scaling on the x-axis can only be relative as it is not possible to quantify the state of the male as one can the length of the female (McFarland & Houston 1981). Other work involving motivational isoclines is discussed in sections 5.7 and 6.3.3. The guppy example, while it illustrates the interaction of internal and external causal factors very neatly, does not tell us anything about the nature of the internal factors except that they vary.

4.3.2   Smooth newts

Newts (*Triturus vulgaris*) have an elaborate courtship in which males display to females before depositing spermatophores that may or may not be picked up by the female's cloaca (Halliday 1974). An encounter between a male and a female consists of a variable number of sequences, each of which contains a period of display followed by a spermatophore transfer phase. During an encounter, the nature of a male's behaviour changes; Fig. 4.5 illustrates schematically these changes for a male that performs three sequences in a row. During the first sequence, the male displays at a high rate and completes each phase of the sequence quickly. By the third sequence, his display rate has fallen markedly, each phase lasts much longer and he tends to oscillate between phases.

These effects have been studied systematically by Halliday (1976). The rationale of this study was that if external causal factors (the behaviour of the female) are kept constant, then any variations in the male's behaviour must be due to processes internal to the male. Female newts can easily be controlled by being held in a harness or 'strait jacket'. Many variations in male behaviour were found to be correlated with one another and, by means of a statistical technique called principal components analysis, it was found that much of the variance in them could be accounted for by a single variable to which the name 'libido' was given. During an encounter, the value of libido declines in each successive sequence. Libido is a statistical construct and the next step was to see whether there was an internal factor which it might represent. The most likely factor is the male's spermatophore supply. At any one time, a male has a limited supply of spermatophores, and as he uses them up, his libido declines.

From the results of this approach, we can suggest that libido explains the relationship between several behavioural measures (display rates, bout lengths, etc.) and a quantifiable internal factor

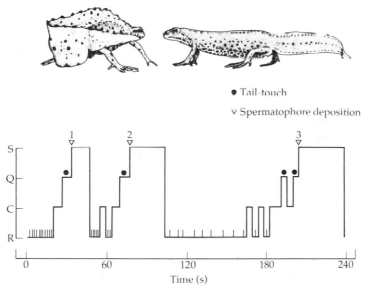

Fig. 4.5. Schematic representation of the sexual behaviour of a male smooth newt (*Triturus vulgaris*) who deposits three spermatophores in the course of a single encounter with a receptive female. The vertical axis shows successive phases in the courtship sequence: R = retreat display, C = creep, Q = quiver, S = spermatophore transfer. The short vertical lines during retreat display represent individual display movements (tail beats). (From Halliday 1976.)

(sperm supply). This is a criterion for the explanatory value of a hypothetical intervening variable (section 4.2.2; Hinde 1982). However, subsequent work on newts raises doubts about the relationship between libido and sperm supply. Male newts court females while 'holding their breath' between ascents to the water surface to breathe. If the amount of oxygen available to a male is altered, his courtship behaviour changes (Halliday 1977). When oxygen is scarce, males display more rapidly and complete each phase of a sequence more quickly; these are effects characteristic of high libido. Conversely, abundant oxygen mimics the effects of low libido. We can rescue libido as a useful explanatory variable by postulating that it is the product of not one but two internal factors, namely sperm supply and oxygen supply (Houston *et al.* 1977). However, even this interpretation is called into question by the following effects observed recently by P.A. Verrell (un-published data).

If a second male is introduced into a tank containing a courting pair, he will often approach the pair and interrupt their courtship, a form of behaviour quite common in newts and salamanders and called sexual interference (Arnold 1976). In response to the presence of an intruding male, the courting male alters his behav-iour in a number of ways. During the display phase he displays at a high rate, as if his libido is increased, but over the whole sequence he takes longer over each phase, as if his libido is reduced. Thus, in the presence of a rival, two aspects of a male's behaviour that are normally positively correlated become negatively correlated. Since it was because these and other scores varied together in a con-sistent way that libido was postulated, the fact that the relationship between two scores can be reversed raises serious doubts about whether libido is useful as an underlying variable.

4.3.3   Rhesus monkeys

As part of an extensive laboratory study of the sexual behaviour of rhesus monkeys (*Macaca mulatta*), Michael and his co-workers have devised a technique for assessing sexual motivation directly (Michael & Bonsall 1977; Michael & Zumpe 1981). To gain access to a sexual partner an animal must press a bar 250 times. The time a female takes to perform this task changes over the course of her menstrual cycle (Fig. 4.6a); she reaches the male fastest near midcycle, when ovulation occurs. This operant technique enables changes in the behaviour of females to be related to changes in the level of sex hormones circulating in their blood (Fig. 4.6b). For example, oestradiol and testosterone show peaks 1–2 days before ovulation, and treatment of ovariectomised females with these

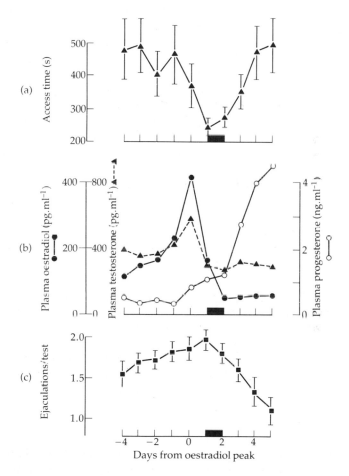

Fig. 4.6. Changes in the behaviour of male and female rhesus monkeys and in hormone levels in plasma of females during the periovulatory period. (a) The mean time taken by females to press a lever 250 times in order to gain access to a male. (b) Changes in three hormones in females. (c) Number of ejaculations achieved by males during a standard mating test. The black bar gives the approximate time of ovulation. (From Michael & Zumpe 1981.)

hormones has shown that they are indeed important in the active interest that females take in males at this time (Keverne 1976).

Other aspects of sexual motivation have also been investigated using this operant procedure, one of them being the degree to which females discriminate between one male and another. They press the bar faster to gain access to certain males, apparently finding them more attractive than others. However, the difference between their speeds of responding for preferred and non-preferred males is greater outside the ovulation period than it is around ovulation (Fig. 4.7; Bonsall *et al.* 1978).

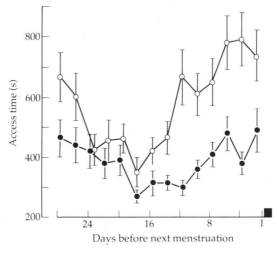

Fig. 4.7. The mean time taken by female rhesus monkeys to press a lever 250 times in order to gain access to preferred (●) and non-preferred (○) males as a function of female reproductive state. (From Bonsall *et al.* 1978.)

Turning to the male, his speed of responding to gain access to a female also peaks at around the time she is ovulating and has been found experimentally to depend on her having oestrogen in her bloodstream. What this oestrogen does is to make her more attractive by stimulating the production of a pheromone, consisting of short-chain aliphatic acids, from her vagina. Females without ovaries (and so lacking oestrogen) can be made attractive either by oestrogen placed directly in the vagina or by introducing a synthetic mixture of these acids there. Not only will a male work harder to reach an attractive female but another measure of his sexual motivation, the number of times he ejaculates during a standard period with her, reaches a maximum when the female is ovulating (Fig. 4.6c).

These three examples illustrate a number of points about the study of motivation. The guppies provide a good illustration of how internal and external causal factors combine to determine the performance of behaviour. An example of the behavioural approach to motivation, this study involved many observations of the responses of males in different reproductive states to females of different sizes. The newt case shows how, again at the behavioural level, observations of behaviour can lead to motivational hypotheses that can be tested experimentally. It also illustrates a common feature of motivational studies, that the more one investigates a particular pattern of behaviour, the more complex are the underlying mechanisms revealed. Finally, the rhesus monkey studies are typical of work in physiological psychology. The animal is placed in an artificial situation where its behaviour is likely to be greatly simplified. This enables the investigator to relate variations in behaviour to changes in physiological state, hormone levels in this example, in a very precise way.

4.4   Sequences of behaviour

Most activities occur in distinct bouts. Thus an animal may feed for a time, preen itself for a while, then attend to the building of its nest, and so on. There are three aspects to the causation of each bout of behaviour; we must consider what causal factors initiate the bout, what factors sustain it while it continues, and what factors cause it to be terminated. Quite different processes may be involved in these three aspects. For example, a bout of feeding might be initiated by the appearance and odour of food, sustained by its flavour perceived in the mouth, and terminated by the distension of the animal's stomach.

For certain activities, there may be little or no variation in behaviour over the course of a bout; a rat drinking or a cow eating grass uses the same motor patterns at the beginning as at the end. In such cases, the principal motivational question to be asked once the behaviour has commenced is what determines the duration of a bout. In contrast, many of the activities with which ethologists are concerned are much more complex and involve a sequence of distinct behaviour patterns, with quite different actions being performed at different stages in a bout. Thus, for a predator a bout of feeding may involve a series of phases, such as stalking, pursuing, killing and eating, each of which has its own characteristic behaviour patterns. Similarly, a bout of sexual behaviour typically involves a complex sequence of interactions between a male and a female, culminating in mating. The causal analysis of such sequences is a major undertaking that usually involves a separate analysis of each constituent phase. The mechanism of switching between behaviour patterns is another aspect that has received considerable study; it is dealt with in section 5.6.

In a behaviour sequence that involves interactions between two individuals, A and B, such as sexual behaviour, transitions by A from one behaviour pattern to another may be causally related both to the behaviour of B and to the previous behaviour of A (Slater 1973). In addition, there may be common causal factors influencing all the behaviour patterns performed in the sequence, and more specific causal factors that terminate the sequence, often referred to as consummatory stimuli (Hinde & Stevenson 1969). It is worth considering the various kinds of causal factors that may be involved.

Where two animals are interacting, sometimes a specific stimulus from one may be more or less essential for the next phase of behaviour from the other. During the sexual behaviour of sticklebacks (*Gasterosteus aculeatus*), the female enters the male's nest and lays her eggs in response to nudges by his snout against the base of her tail. The female then pushes her way out of the nest and this provides the stimulus for him to follow her and fertilise the eggs (Tinbergen 1951). During newt courtship, the male deposits a spermatophore in response to the female touching his tail with her snout (Halliday 1975). This is not an absolutely stereotyped interaction, however. Males with many spermatophores sometimes deposit one without receiving a tail-touch and males that have depleted their spermatophore supply may require as many as four tail-touches to elicit the deposition of the last one (Halliday 1976). Once again, we find that a response to an external stimulus is modulated by an internal causal factor.

External stimuli which do not emanate from other animals may also have such specific effects. The completion of one phase in a complex behaviour sequence may provide the stimulus that initiates the next phase. In building her nest, a female canary (*Serinus canaria*) must gather nest material, carry it to her nest site and then arrange it in an appropriate manner. Gathering of nest material tends to elicit carrying, which in turn elicits building (Hinde 1958). Similarly, the switch from one phase of nest digging to another by digger wasps (*Sphex ichneumoneus*) is elicited by completion of the previous phase (Brockmann 1980). In some cases like this the switch from one action to the next may be for comparatively trivial reasons. A female canary cannot continue carrying material to her nest once she has reached it and, furthermore, she has then placed herself in exactly the right stimulus situation for building the material into a nest to commence. In many cases, however, internal changes are involved as well. A female dove will sit on her eggs when she has laid them, but will also adopt eggs placed in her nest shortly before this. She is not totally flexible, however. She will ignore a nest and eggs given to her when she is first paired with a male (Lehrman 1961); her internal state is not yet ready for incubation (see section 4.5.1).

Although the performance of particular behaviour patterns within a sequence may be dependent on specific factors, they may also be influenced by factors common to all phases. Thus, a certain level of reproductive hormones may be necessary for all activities involved in courtship or nest building to occur. As we saw in section 4.3.2, the amount of sperm that a male newt has available has an effect on all phases of his courtship sequence. Within a sequence, however, different phases may vary in the way they are influenced by a common causal factor. In nest building by canaries, gathering, carrying and arranging of nest material are all influenced by daylength and by the sex hormone oestradiol (Hinde & Steel 1978). Early in the season, when oestradiol is low, a female gathers material but does not carry it; later she gathers it but carries it in a haphazard fashion; only when in full reproductive condition and when days are long does she perform complete nest building sequences. Injection of oestradiol early in the season elicits nest building activities and the higher the dose, the more likely it is that complete nest building sequences will be performed, but daylength is important too. The dose of oestradiol has to be near lethal to elicit nest building in birds kept on short days. Here, then, two different common causal factors affect a whole sequence and there is a subtle interaction between them.

In addition to the specific and general causal factors which influence their course, complex behaviour sequences often have a recognisable functional consequence or 'goal', such as the fertilisation of eggs or the completion of a nest. External stimuli that arise as a consequence of the completion of a sequence may have the effect of terminating the sequence, just as one phase in the sequence may lead to stimuli appropriate to the next. These are referred to as consummatory stimuli. When a male stickleback has fertilised the eggs his behaviour changes dramatically; from courting the female he attacks her and drives her away. It is not the act of spawning that causes this switch, but the odour of newly laid eggs in his nest (Sevenster-Bol 1962). Surprisingly, the successful transfer of a spermatophore to a female does not act as a consummatory stimulus for male newts (Halliday 1976). Rather, courtship appears to be terminated by a combination of a male's sperm supply becoming depleted and his need to ascend to the water surface to breathe (Halliday & Sweatman 1976). Here, as with the situation that terminates feeding or the ejaculation which leads to a period of sexual quiescence, the factors involved in terminating a sequence are primarily internal. However, even in apparently clear-cut cases, external stimuli have been found to have a role as well. A novel taste or smell can elicit more eating from a previously satiated animal and, in certain species, a male which has apparently copulated to exhaustion may commence to mate again if presented with a new female (the Coolidge effect).

Not all patterns of behaviour are subject to causal factors of the sorts just discussed. Some activities seem to occur largely independently of external stimuli. An example is the 'creeping through' shown by male sticklebacks ('t Hart 1978). In the period between the completion of his nest and the appearance of a female, the male periodically pushes his way through the nest, apparently keeping the tunnel through it clear. If a nest is experimentally shortened or a hole is made in its roof out of which the male has to swim, the duration of each bout of creeping-through becomes reduced. The male's response to this is to shorten the interval before he next creeps through his nest. On subsequent visits he may stay longer inside the nest, and he sometimes rebuilds it to its previous size. Conversely, artificially lengthening the nest causes him to break out of it when creeping-through has lasted for the normal time. Preventing a male from leaving his nest after creeping through it does not extend the interval before his next visit. These results suggest that male sticklebacks normally creep through their nests according to a fixed schedule, with bouts of creeping-through lasting for a specified duration and separated by intervals the length of which is determined by that duration. There is no evidence to suggest that this behaviour is altered in response to the effect that it has on the nest.

4.5   General factors in motivation

In most modern accounts of motivation, causal factors are seen as having, to varying degrees, specific effects on specific activities. In the past, by contrast, it has sometimes been suggested that certain causal factors may act to influence some general drive whose level in turn affects all categories of behaviour (see also section 5.4). The general-drive concept is now widely rejected (Bolles 1975; Hinde 1970). However, there remains the common observation that often animals are more alert and responsive at some times than at others.

A general factor, usually referred to as *arousal*, has sometimes been proposed in order to account for such changes. The exact hypothesis may take several different forms (Andrew 1974). Arousal may be seen simply as a correlate of neural activity in the reticular activating system of the brain, although this says nothing of its postulated effects on behaviour. The behavioural implications are most often one or other of two types, in both of which the level of arousal is seen as varying according to whether the animal is unconscious, asleep, resting or alert. In the first, arousal is thought to affect the responsiveness of the animal to a wide variety of stimuli such as food, predators and sexual partners. It therefore influences a number of quite different motivational systems. At a rather general, and trivial, level there is obviously some truth in this; animals *are* more responsive when awake than when asleep. However, the evidence for a continuum of different levels of responsiveness in waking life is less compelling.

The other common form of the arousal hypothesis is expressed in terms of activation. This proposes that a general variable affects which of a number of potential behaviour patterns is shown and might, for example, account for the way in which different activities fit into the circadian rhythms that animals show (Brady 1975;

section 4.5.2). Male zebra finches (*Taeniopygia guttata*) oscillate between periods of high activity and feeding and periods in which preening and rest are the main behaviours shown, with song tending to occur between the two. This is true both over the course of a day (the birds are more active in the morning than in the afternoon) and over much shorter time periods, and Ollason and Slater (1973) suggested that these relationships might be accounted for by fluctuations in a single activation variable. Experiments by Slater and Wood (1977) did not, however, support this idea as it proved possible to dissociate the rhythms of the various activities involved using short-term cycles of light intensity. At this stage, therefore, the evidence is against arousal-like variables having a major influence on behaviour, whether they affect responsiveness or the activation of particular behaviour patterns.

It is the case, however, that the performance of some behaviour patterns can be enhanced by factors that appear quite irrelevant to them. Frustration is one factor which has rather generalised effects on behaviour (section 5.4.3). More specifically, electric shocks can facilitate sexual behaviour in male rats (Barfield & Sachs 1968) and a wide variety of activities can be facilitated in rats by pinching their tails (see Roper 1980). Such effects, reported from highly unnatural situations, are hard to interpret; one might have expected shocks and tail pinches to elicit avoidance behaviour that would compete with, not facilitate, other activities.

While the role in the motivation of behaviour of factors with general effects, such as arousal and noxious stimulation, remains uncertain, it is clear that certain physiological phenomena, whose role can be defined, have a facilitating effect on a variety of activities. We shall consider two such phenomena, reproductive hormones and endogenous rhythms.

4.5.1   Hormones and behaviour

In the context of this chapter, it is possible to give only the most superficial account of what is known of the role of hormones in motivation. Of all the internal factors that influence behaviour, hormones are the most fully understood. There are two reasons for this. First, hormone levels in the body can be precisely measured and related to behavioural changes (as in Fig. 4.6). Secondly, and more significantly, the endocrine organs that secrete hormones can be removed with little damage to the animal. It is then possible to examine the role of a specific hormone by injecting known quantities of it, in effect mimicking the secretion of the removed organ. For example, the role of testosterone in the behaviour of male animals is commonly studied by injecting it into castrated individuals.

Hormones are general factors in motivation in two senses. First, they typically influence several aspects of behaviour and may be involved in the causation of a number of different activities. An obvious example is testosterone, which influences not only sexual behaviour but also aggression. It appears that testosterone may have other effects. In chicks and mice, it increases the persistence with which individuals search for a particular type of food (Andrew 1976). In courtship also, some of the effects of testosterone may be because it increases the male's persistence. The second sense in which hormones have general effects is that they influence behaviour over a relatively long time scale. Reproductive hormones are generally necessary for the expression of sexual behaviour, but levels of these hormones do not change from moment to moment, so that changes in sexual behaviour cannot be attributed to concurrent changes in hormone levels but must be explained in terms of more immediate events, such as the behaviour of a mate. The time-scale over which most hormones act on behaviour may be an animal's lifetime, as in sexual maturation, one or more years, as in animals that breed more or less annually, or a matter of days or weeks, as in the control of reproductive cycles in mammals. The 'fight or flight' hormone adrenalin, which brings about the effects one feels when suddenly angered, is unusual in having effects on behaviour over a matter of minutes or even seconds, but future research may well reveal other such short-term influences of hormones on behaviour.

Although the effects of hormones on motivation are obviously internal, it is important to stress that they do not work in isolation but interact in complex ways with causal factors from the outside. Hormone secretion is commonly itself dependent on external stimuli; for instance, in male birds testosterone secretion starts in spring in response to increased daylength. Also, a hormone cannot usually elicit behaviour on its own without appropriate external stimuli, such as those provided by a mate. One of the most elegant features of the relationship between hormones and behav-

Chapter 4

iour is the way in which the hormones appropriate to an aspect of behaviour have often been found to be secreted in advance of the situation in which that behaviour appears. This is a good example of a feedforward mechanism (see section 6.2.3).

One of the most complete analyses of the interaction between

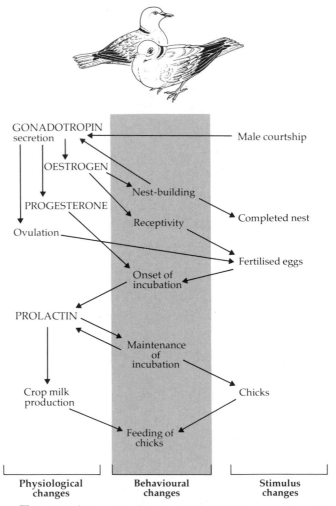

Fig. 4.8. The principal interactions between physiological changes, external stimuli and behaviour involved in the reproductive behaviour of female ring doves. Hormones are shown in capitals. (From Slater 1978a.)

hormones, external stimuli and behaviour is that which has been carried out on the reproductive behaviour of doves (Fig. 4.8). Originally started by D.S. Lehrman, this work has been carried on by many others (see summaries by Silver (1978) and by Cheng (1979)). If a pair of Barbary doves (*Streptopelia risoria*) are housed together with some nest material and a bowl in which to build a nest, they will go through a week-long courtship period, towards the end of which they mate and build a nest. Two eggs are then laid and, until these hatch about two weeks later, male and female share in incubating them. By the time hatching occurs, the crops of both parents have developed the capacity to secrete a milky fluid on which the nestlings, called squabs, are fed.

As shown in Fig. 4.8, the behaviour of a female dove is influenced by a number of physiological factors, including at least four different hormones, and by external stimuli from her mate, the nest, eggs and squabs. A feature of this system is that at each stage the female is stimulated to secrete the hormones that prepare her for the next stage. Male courtship stimulates the secretion of oestrogen and progesterone, which cause her to become sexually receptive and which are necessary for nest building and incubation. Incubation in both sexes stimulates prolactin secretion and this hormone is necessary both for incubation to be maintained and for the secretion of crop milk. That prolactin secretion is stimulated by and stimulates incubation indicates positive feedback between the behaviour and a causal factor that underlies it (see section 6.2.2). This mechanism ensures not only that incubation is maintained but also that when it ends and the eggs hatch, the birds are secreting crop milk and are thus ready to feed the young immediately.

The reproductive behaviour of male doves appears to be controlled in a rather different way from that of females, although prolactin has similar effects on both sexes. Whereas it is clear that testosterone, which shows a marked rise after pair formation, is involved, many of the changes in a male's behaviour are primarily responses to the female's activities, not to his own hormonal state. For example, he begins nest building in response to the female starting to spend long periods at the nest site, not to any specific endocrine events (Erickson & Martinez-Vargas 1975). It appears that testosterone is involved in the causation of a wide range of male activities, including several different courtship displays, nest building and incubation, but that their precise coordination is determined by external cues from the female, the nest and its contents. Perhaps what this example illustrates most clearly is that it is a mistake to think of causal factors as either specific or general because some of them are more limited in their effects than others. Testosterone does not affect all activities performed by male doves, nor are its effects limited to just one of them, but it does influence a variety of activities to varying degrees.

4.5.2   Endogenous rhythms

In most animals, some activities are cyclical in their timing. Rhythms of behaviour may have a periodicity related to the year (circannual rhythms), the lunar cycle (lunar or tidal rhythms) or the day (circadian rhythms). Some are less obviously related to environmental cycles. In most instances, behavioural rhythmicity can be explained in adaptive terms, animals typically performing an activity at those times when it is best to do so (Daan 1981). Behavioural rhythms may simply reflect rhythmic changes in relevant environmental factors such as food availability, but in many instances they are controlled by internal or endogenous rhythms which cycle largely independently of external cues (Brady 1979, 1982). An example of a circannual rhythm in behaviour is the migratory restlessness shown by many birds. At the two appropriate times of year, captive migrants kept on a constant daylength jump and flutter around their cages at night. Migratory restlessness is commonly also influenced by changes in daylength but, in some species, these are not necessary for it to occur at around the right time of year (Gwinner 1981). The fact that these cycles are not strictly in phase with those in the outside world indicates that they are truly endogenous rather than simply influenced by external factors which have not been controlled.

There are many animals that show circadian rhythms of feeding behaviour in the laboratory, where food is usually available at all times (Toates 1979b), as well as in the field, where food availability often fluctuates over the course of a day. Daily patterns of feeding can be understood in terms of an interaction between an animal's food requirements and fluctuations in food availability in its environment (Daan 1981). For example, kestrels (*Falco tinnunculus*) hunt for rodents at times when they are most available

(Rijnsdorp *et al.* 1981; Fig. 4.9). Much of the prey caught is not eaten immediately but is cached (hidden), especially during the middle of the day, and is retrieved at times when food is scarce. Thus, much cached food is retrieved just before nightfall, enabling a bird to fill its stomach for the night. In some species, rhythmic aspects of feeding are associated with specific dietary needs. For example, domestic fowl (*Gallus domesticus*) show an increase in their intake of calcium-rich food in the few hours prior to ovulation (Hughes 1972). Calcium is essential for the production of egg shells. Rabbits (*Oryctolagus cuniculus*) re-ingest faecal pellets just before dawn, and this behaviour is controlled by an endogenous circadian rhythm (Hörnicke & Batsch 1977). Faecal pellets are rich in symbiotic microorganisms that break down cellulose in the gut and those eaten before dawn are stored to facilitate the efficient digestion of the next night's food intake.

On a much shorter time-scale, an endogenous rhythm has been implicated in the feeding behaviour of fifth-instar locusts (*Locusta migratoria*) (Simpson 1981, 1982). These animals eat in meals of varying length, separated by variable time intervals, and they feed

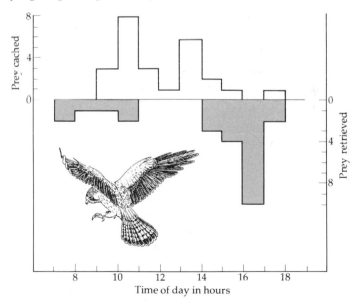

Fig. 4.9. The frequency of caching (upper histogram) and retrieval of cached prey (lower histogram) by kestrels, as a function of time of day. (From Daan 1981.)

more in the light than in the dark period. The onset of feeding bouts coincides with the peaks in a sinusoidal rhythm, whose period is constant within an individual but varies from 12.0 to 16.5 min between individuals (Fig. 4.10a). Locusts do not feed at every peak, but other active behaviour patterns, such as locomotion, tend to take place at peaks when feeding does not occur. The interval between feeding bouts is partly dependent on the size of the previous meal; the larger a meal, the longer is the interval before the next begins. This suggests that the onset of feeding is

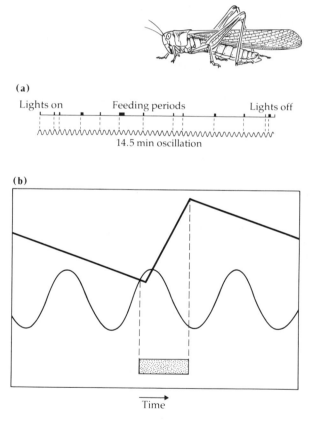

(a)

Lights on   Feeding periods   Lights off

14.5 min oscillation

(b)

Time

Fig. 4.10. (a) The feeding record of a single fifth-instar locust over a 12-hour period superimposed on an oscillation with a period of 14.5 min. (From Simpson 1981.) (b) Hypothetical interaction between a feeding threshold (heavy line) and an oscillating rhythm (thin line) causing a feeding bout (shaded bar).

determined by the interaction between a feeding threshold, which falls with time since the last meal, and the underlying oscillation (Fig. 4.10b).

Sleep is an especially interesting example of a behaviour which normally occurs rhythmically. The sleep requirements of different animal species vary enormously with some, such as dolphins and shrews, never having been recorded as sleeping, while others, like the sloth, may sleep for as much as 20 hours per day. Many physiological functions occur predominantly during sleep, but Meddis (1975) argues that sleep did not evolve because of this; he suggests instead that it is primarily a mechanism to keep animals 'out of harm's way' at times of day when they are ill-adapted to function, as, for example, is the case for many animals when it is dark. As anyone who has suffered from jet lag will realise, the timing of sleep is determined both by changes in the outside world and by an endogenous circadian rhythm which must be re-set when one moves from one time zone to another.

If an endogenous rhythm is involved in the causation of an activity, then the motivation of that activity is not simply dependent on the physiological processes that the activity affects. Thus, an external causal factor of a given cue strength may elicit a behaviour pattern at one time of day but not at another, even though the animal's internal state is the same. In the case of feeding, this means that an animal may apparently be 'hungry', not simply because it is experiencing a nutrient deficit, but because it is that time of day when it normally eats. Once again, we see that it is often too simplistic to consider the motivation of an activity as being due to a single underlying process.

4.6   Diversity in motivational processes

Lorenz's hydraulic model was not specifically intended for one type of behaviour rather than another, and it has often been assumed that all behaviour patterns have essentially the same kind of motivational mechanisms underlying them. It is now clear that such an assumption is untenable and that different activities are governed by very different motivational processes. Moreover, even when considering the same activity, there are differences in the motivational mechanisms found in different species.

An important determinant of the motivational processes

underlying a particular activity is the function that the activity plays in the animal's life. For example, a basic dichotomy can be made between those activities that are involved in homeostasis and those that are not (Hogan 1980). Homeostasis (literally, 'staying the same') is the phenomenon that animals maintain various physiological variables at an approximately constant level. An obvious example is body temperature in warm-blooded vertebrates such as ourselves. Homeostatic systems are typically controlled by negative feedback mechanisms (see section 6.2.2). If its temperature rises above or falls below its normal value, an animal will show physiological and behavioural responses that counteract the disturbance. Activities that clearly have a role in homeostasis are feeding and drinking, which maintain nutrients and water within the body at levels necessary for an animal's survival. Both act to replace vital commodities that are constantly being expended or lost. By contrast, activities such as aggression, sexual behaviour, exploration and play can be regarded as non-homeostatic, as their performance does not serve to correct departures from a physiological norm. Indeed, it is generally the case that the performance of such activities is best understood in terms of changes in the environment. For example, the amount of aggression shown in the laboratory by animals such as rats and mice can be manipulated by varying such factors as the level of crowding and the number of refuges available (Archer 1970), and exploration can be elicited by opening up an unfamiliar portion of a complex maze (Barnett & Cowan 1976).

In some respects, aggression and exploration can be viewed in homeostatic terms (see Archer 1976; Toates 1980a), but the variable controlled is not a physiological one like temperature or blood glucose. In a territorial animal, for example, aggression can be seen as occurring in response to departures from the norm that there are no intruders in the territory, and exploration in response to changes in the environment resulting in a discrepancy between the animal's knowledge of its home range and the norm that it is familiar with it. However, this way of expressing things simply shows that, as with sexual behaviour (section 6.5), some of the discussion of these motivational systems can be phrased in terms of control theory. It does not imply that there are strong similarities in the exact mechanisms involved.

An example of good evidence that different activities may be controlled by different motivational processes, according to the degree to which they are involved in maintaining homeostasis, comes from studies of Siamese fighting fish (*Betta splendens*) (Hogan *et al.* 1970). Fighting fish were trained to swim through a small tunnel to obtain a reward, either food or an opportunity to display aggressively to a mirror. At first, a fish received a reward for every correct swimming-through response, but gradually the ratio of responses to rewards was changed until six responses were required to gain each reward (Fig. 4.11). When the reward was food, fish increased the number of their responses so that, over the course of a 12-hour test session, they maintained the same level of food intake. In contrast, when the opportunity to behave aggressively was the reward, they maintained a more or less constant number of responses and so experienced a lower reward rate.

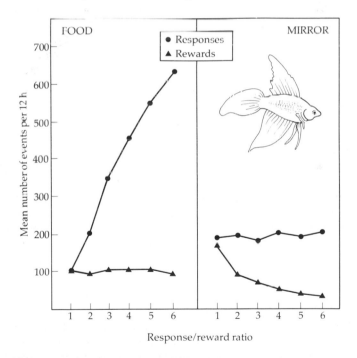

Fig. 4.11. Changes in the number of responses shown and rewards received by Siamese fighting fish, as a function of changing response/reward ratio, when the reward is food (left) or the opportunity to behave aggressively to a mirror (right). (From Hogan *et al.* 1970.)

These results are consistent with the view that feeding fulfils a regulatory function and that aggression does not. When obtaining food becomes more costly, a fish works harder so as to regulate its food intake. Its aggressive behaviour shows no such effect.

This is a particularly clear-cut example and, with others, makes the point that each motivational system must be looked at individually without preconceptions about the factors affecting it. Feeding and drinking have a clear homeostatic function and, because of this, must take place frequently and fairly regularly. Sexual behaviour is not regulatory and, although it is subject to exhaustion as we have seen in the case of newts, the other factors affecting it have little parallel with feeding or drinking. Animals can survive without sex; indeed, they often go for long periods without it then copulate several times in quick succession. In female mammals its occurrence is usually affected by oestrous cycles; in males copulation may actually potentiate further activity to some degree. A male rat that has not mated for some time needs fewer intromissions to achieve a second ejaculation than he did for his first (Beach & Whalen 1959). Aggressive behaviour, like sex, does not have to be shown for an animal to survive, but is typically used by animals when they need it, for example to gain some contested resource. It is influenced by physiological factors such as hormones but the factors which trigger it are mainly in the external world. The evidence is strongly against the idea that aggression builds up with time until it must be expressed, as Lorenz (1966) argued in his influential book, presumably with his hydraulic model in mind. One of the fascinating things about motivational systems is that they differ from one another to match the different needs that they meet. Natural selection has tailored them to meet the exact requirements of the animals that they serve.

4.6.1   *Behavioural regulation*

For activities such as feeding and drinking, models based on the idea that behaviour serves to regulate some physiological variable, such as blood sugar or body weight in the case of feeding and body water in the case of drinking, have been very useful in furthering our understanding of motivation (Chapter 6). It is important, however, to consider a number of limitations in their application.

(1) The fact that a variable remains reasonably constant over

time cannot be taken as conclusive evidence that it is being regulated by behaviour. For example, an animal's food intake may be determined more by the availability of food in its environment than by its homeostatic requirements. As Bolles (1980) vividly puts it, a cat living where there are a lot of mice will be a fat cat, whereas one living where mice are scarce will be scrawny. There is no reason to suppose that one is adjusting its behaviour to maintain a high body weight, the other to maintain a low one.

(2) A physiological variable may indeed be regulated, but mainly by mechanisms other than behaviour. The concentration of an animal's body fluids is influenced not only by how much water it drinks but also by its kidneys, which are very efficient at conserving water or at excreting excess water, as circumstances demand. We should not expect, therefore, that its drinking behaviour is exactly matched to its water requirements. It may, for example, drink so as to exceed its requirements, and then achieve regulation of its body fluids by excreting the surplus through the kidneys.

(3) A physiological variable may be regulated not over a matter of minutes or hours but in the longer term. When an animal eats, several physiological changes are initiated that occur over different time scales. Thus, the animal's stomach becomes distended during the course of a meal, but levels of glucose in the blood and of glycogen in the liver change more slowly and will continue to do so long after the meal has finished, as digestion proceeds in the gut. If a meal is to be terminated by negative feedback from its physiological consequences, the particular consequences involved must be fairly immediate. In fact, it is clear that immediate feedback, for example from the stomach, plays a relatively minor role in the control of meal size (Toates 1980a, and Chapter 6). Experiments by Booth (1972) provide evidence that rats adjust the size of their meals through learning, on the basis of slow-acting feedback processes, a phenomenon called conditioned satiety. It appears that they associate specific flavours with the metabolic consequences of the food they eat. If a rat eats a food with a particular flavour and subsequently detects a low nutritional input, it learns to eat more of that food, recognising it on subsequent occasions by its flavour. Conversely, rats fed on food with a high nutritive content will learn to eat small meals. It seems then, that a meal is not terminated simply by immediate negative feedback but, at least in part, by a

rat's ability to anticipate when a meal has been large enough to fulfil its needs. These results, as well as emphasising that we must consider long-term as well as immediate feedback processes, make the point that one cannot draw a rigid distinction between motivation and the influence of learning (section 4.2.1). Learning may also be involved in the feeding behaviour of kestrels, which kill more prey than they eat immediately, in apparent anticipation of their future needs (section 4.5.2).

(4) Animals may regulate aspects of their physiology at some times but not at others. Many animals maintain a relatively stable body weight for most of the year but put on weight prior to periods of high energetic demand such as migration or hibernation, or lose weight when they become engaged in time-consuming activities such as breeding. Junglefowl hens (*Gallus gallus spadiceus*), for example, drastically reduce their food intake and suffer a consequent marked weight loss while incubating their eggs, even if food is supplied close to the nest (Sherry *et al.* 1980). Such marked changes in behaviour suggest that, if feeding behaviour is based on physiological regulation, the value of the variable that is being maintained (its set point) may be changed according to demands imposed by other activities. It may also be the case that animals behave in a regulatory fashion only when their environment allows them to. For example, mice can be trained to press a disc to obtain nest material, and the number of responses required to gain a reward can be varied (Roper 1973, 1975). If necessary, they will increase the rate at which they work for nest material so as to obtain a fairly constant amount, but only if the nest material dispenser is close (3.5 cm) to the disc. If disc and dispenser are separated by 6 cm, they will not increase their work rate and so fail to maintain their nest material supply.

Phenomena such as these may have the effect of producing confounding variations in what may otherwise appear to be a simple regulatory system. While negative feedback mechanisms are incorporated in many models of motivation (see Chapter 6), it is clear that the concept of physiological regulation is inadequate for explaining many aspects of motivation. In a very thorough review of the literature on motivation, Bolles (1975) wrote:

> 'The idea that an animal's motivation reflects its need for this or that commodity in order to establish homeostatic equilib-
> rium has only occasionally been demonstrated experiment-

ally. The idea that an animal's motivation is an automatic adjustment to its state of need is attractive and appealing, but is not justified by the facts.'

4.7 Conclusion

This chapter will have served a useful purpose if it has convinced the reader that there are no unifying principles or widely accepted general theories in the study of motivation. There is probably less consensus about the nature of motivational mechanisms than there is in any other area of ethology. The kinds of models and hypotheses adopted by an individual researcher depend on a variety of factors, including the kind of behaviour they are investigating, the species they are working with and the intellectual background in which they have been trained. While this diversity of approach is confusing to those studying motivation for the first time, it is no bad thing. It is clear that animals show enormous diversity in the motivational mechanisms that they possess and it is only through the synthesis and integration of a variety of approaches that we are ever likely to understand those mechanisms fully.

4.8 Selected reading

A very thorough and clear account of motivation is provided in *Animal Behaviour, A Systems Approach* by Toates (1980a), a book which considers the subject from both ethological and psychological standpoints, devoting separate chapters to different activities such as feeding, aggression and sex. The psychological literature is thoroughly and critically reviewed in *Theory of Motivation* by Bolles (1975). An attempt to bridge the gap between ethology and psychology, and to explore their common ground, is the book *Analysis of Motivational Processes,* edited by Toates and Halliday (1980). Particularly recommended in this book are the chapter by Hogan on the concept of homeostasis and that by Bolles which, in a very forceful way, shifts the idea of regulation from the internal environment to the outside world.

CHAPTER 5
INTERACTIONS
BETWEEN
ACTIVITIES

R.H. McCLEERY

5.1   Introduction

A striking thing about the behaviour of animals is the orderly way in which they go about their business. When we consider that at any instant an animal is subject to a variety of internal and external stimuli and conditions which predispose it to perform a variety of incompatible activities, this superficial order becomes even more remarkable. There are two aspects to the problem, both of which are considered in this chapter. First, there are clearly causal mechanisms of interaction between the conflicting pressures on the animal which lead to the appearance of organisation in behaviour. Secondly, the orderly patterning of behaviour presents an appearance of purpose and of established priorities between activities. One can think of the animal making decisions about which activity to perform. These causal and functional aspects are treated more or less separately here, except that I shall try to show how this particular view of causation leads naturally to a potentially powerful way of looking at function.

One of the things that makes the task of understanding interactions between activities more difficult than it need be is the plethora of theories of behaviour compared to the relative scarcity of information which clearly allows one to decide amongst them. Peters (1960) traces this problem back at least as far as Hobbes' theory of human action, and points out that many theories have been instigated more by a desire for a complete theory of behaviour than by puzzlement about concrete problems in psychology or ethology. The result of this rather grandiose habit of thought has been such enormities (Peters 1960) as Hull's *Principles of Behavior*

(1943), impeccably hypothetico-deductive but extraordinarily unsuccessful at explaining real behaviour. Such matters would be of purely historical interest but for the fact that when systems like this are finally swept away by their lack of contact with reality they leave behind them a wrack of theoretical terms which may have wide currency but which can be positively misleading outside their original context. In places I have therefore looked in some detail at views of motivation which may be considered outmoded but which continue to influence current thinking through the survival of some of their concepts.

5.1.1   Classifying behavioural elements

The stream of behaviour is continuous but to say anything useful about it some means of breaking it up into units must be adopted. Usually this is seen as a methodological or statistical problem (see section 3.2) and has been likened to that of the nineteenth century taxonomist classifying museum specimens (Slater 1978). The analogy is perhaps more profound than intended, for just as taxonomy can be seen as a search for natural relationships which reflect similarity by descent rather than simply morphological congruence, so the classification of behaviour can be seen as a search for 'true' underlying relatedness between components. It is sometimes demanded that units be defined statistically and labelled with neutral labels (Slater 1978b) but this procedure may obscure important relationships. Hinde (1970) calls an alternative classification 'description by consequence' as in 'picking up nest material' or 'approaching'; here the descriptions cover all patterns which lead (or could lead) to a specified result. The two forms of classification reflect an implicit belief that some activities are related to each other either through common machinery or through functional association, or both. Hence the elements of 'aggressive behaviour' are grouped together both because all appear to be determined by similar causal factors and because all have similar consequences.

The grouping of elements of behaviour presents the appearance of a hierarchical organisation (see section 3.4; Dawkins 1976). There does not seem to be any generally agreed terminology for such hierarchies. Figure 5.1 shows a scheme based loosely on that of McFarland and Sibly (1975) which will be used for discussion in this chapter. Major components of behaviour such as feeding or courtship are called 'activities' and are themselves made up of 'actions' such as 'eating' or 'searching'. McFarland and Sibly (1975) define an action as being 'an identifiable pattern of behaviour such that it is always possible to decide whether it has occurred or not; an action cannot occur at the same time as another'. Actions have 'characteristics' such as 'pecking' and 'swallowing' and may be considered to be related to each other if they have characteristics in common (cf. 'eat' and 'reject' in Fig. 5.1). It is debatable whether the actions comprising activities are unique to those activities or whether the same action may occur in more than one activity. For example in Fig. 5.1 'searching' appears as an action in the activities 'feeding' and 'courtship'. This seems a natural idea, but it does mean that if one had only a brief glimpse of the animal then it might not be possible to say which activity it was pursuing; in this case it might be either feeding or courting. McFarland and Sibly (1975) argue that such an action should be differentiated into 'food search' or 'mate search' or, if that is impossible, simply ignored. Ignoring it leads to the paradox of having undoubted actions which do not form a part of any activity, but admitting actions to more than one activity leads, strictly speaking, to the equally paradoxical view that the animal is doing several things at once. In this case the animal would be considered to be both feeding and courting. The answer is that the correctness of a classification of behaviour depends on what it is needed for. Labels are clearly needed for the different levels of the behavioural hierarchy and I

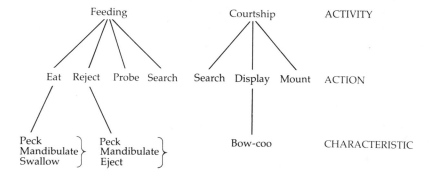

Fig. 5.1. Terminological scheme for hierarchical organisation in behaviour. (Based on McFarland & Sibly 1975.)

shall use the terminology of Fig. 5.1, albeit rather loosely and without enforcing the 'non-overlapping' requirement.

5.2   The mechanical mollusc—a success story

There are two extreme kinds of observation that can be made in order to find out about motivational mechanisms. On the one hand the animal can be treated as a 'black box' and an attempt can be made to deduce the nature of the internal processes from the input–output relationships. Unfortunately, any given set of responses to various inputs could be generated by a variety of internal mechanisms. At the other extreme, attention can be confined to direct investigation of the nervous system, but when we start trying to understand the overall functioning of an animal's nervous system by inserting electrodes or drugs or removing parts of its brain we are in a similar position to someone trying to understand a computer using only an oscilloscope and a pair of wire-cutters (Oatley 1978). The best approach is obviously intermediate between these extremes, and usually starts with behavioural observations; in favourable cases it may be possible to proceed to the direct observation of internal mechanisms.

One of the most successful attempts to understand the interactions between behavioural systems from the level of overt behaviour down to the neuronal mechanisms governing priorities has been the work of W.J. Davis and his colleagues on the marine gastropod *Pleurobranchaea californica* (Davis 1976). *Pleurobranchaea* is an opportunistic carnivore and feeds on dead organisms, the eggs and adults of the related gastropod *Aplysia,* and other *Pleurobranchaea*. Its behavioural repertoire consists of relatively few activities, organised in a largely hierarchical way as shown in Fig. 5.2. At the top of the hierarchy, escape always suppresses all other activities. Egg laying suppresses feeding, but can occur simultaneously with righting and probably with mating, though this has not been demonstrated in *Pleurobranchaea*. Withdrawal of the oral veil due to mechanical stimulation can also occur simultaneously with righting or mating, but is suppressed by feeding. However, when there are few food stimuli or the animal is nearly satiated, withdrawal from tactile stimulation suppresses feeding, so the feeding/withdrawal relationship is one of reciprocal inhibition (Kovac & Davis 1980a).

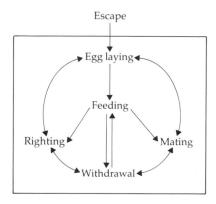

A —→ B	A dominates B
A ←→ B	A and B can be simultaneous
A ⇌ B	Reciprocal inhibition between A and B

Fig. 5.2. Behavioural hierarchy in *Pleurobranchaea californica.* The bidirectional arrow between egg laying and mating represents a relationship which has been demonstrated for *Phyllaplysia* but not *Pleurobranchaea.* (From Kovac & Davis 1980a.)

Figure 5.2 is based on entirely behavioural observations, but several of the links have been investigated physiologically. The suppression of feeding by egg laying is due to a hormone that has been shown to bind to sites in the buccal ganglion, which is known to control feeding, though the precise nature of the hormone's action is not yet clear. The different experiences which reduce feeding rate (satiety and aversive stimulation) have been shown to exert synaptic inhibition on the feeding command neuron, a single neuron which when stimulated electrically causes feeding to begin. The inhibition of veil withdrawal by feeding has been found to be mediated by a pair of neurons that are active during feeding and that, when stimulated, abolish the motor output to the veil which is usually produced when it is mechanically stimulated. These relationships hold in an isolated nervous system where sensory inputs are provided by direct stimulation of the stumps of the sensory nerves, so it can safely be concluded that purely central inhibitory mechanisms are involved (Kovac & Davis 1977). In an intact animal, righting is suppressed by a food stimulus even when no actual feeding occurs because the animal is satiated. This implies that chemosensory inputs which initiate feeding also have a direct inhibitory effect on righting. Satiating the animal abolishes the inhibition of withdrawal normally caused by a food stimulus, implying a direct inhibitory connection between the feeding command neurons and the central neurons controlling withdrawal (Davis *et al.* 1977).

These arrangements appear to make functional sense. Since *Pleurobranchaea* is an opportunist, it is important for it to grab food whenever food happens to be available, unless there is imminent danger of predation, in which case the most important thing to do is escape. The herbivorous but otherwise similar *Aplysia* gives feeding a lower priority, and its responsiveness to food shows a circadian fluctuation. In *Pleurobranchaea* the inhibition of feeding by egg laying appears to be a solution to the problem that the animal will readily eat its own eggs in the absence of the hormone, since it seems unable to distinguish them from those of other *Pleurobranchaea* or of *Aplysia*.

In this one animal, therefore, are found examples of a fixed behavioural hierarchy, of a modifiable hierarchy and of non-hierarchical reciprocal inhibition, and sets of compatible and incompatible activities. In addition, a considerable degree of understanding of the neural mechanisms and some understanding of their functional significance has been achieved. However it is worth asking at this point whether one should talk of motivation at all in an animal such as *Pleurobranchaea* when we have a pretty good understanding of the neural mechanisms governing its behaviour. Would motivational constructs become redundant if we had a complete wiring diagram and performance specification of the nervous system? In fact such a 'neurophysiologist's nirvana' (Dawkins 1976) would not constitute an understanding of behaviour in any real sense. Dethier (1966), in a closely argued discussion, points out that the issue is not so much whether motivational concepts improve our understanding of the animal's behaviour (in Dethier's case that of the blowfly), as whether the processes involved in such animals are anything like those which occur in animals where we have little or no knowledge of the internal wiring. In terms of common ancestry, the blowfly and *Pleurobranchaea* clearly have little in common with vertebrates, but there is a level at which generalisations about common mechanisms are valid; for example the functioning of axons and synapses are homologous in insects and man in much the same way as mitochondria or chromosomes are homologous. If phenomena such as learning and motivation are emergent properties of aggregates of neurons, then motivation in higher vertebrates may involve no new principles over similar phenomena in insects or molluscs. The processes can be considered homologous in the sense that they are functions of homologous entities, albeit in a more elaborate form. Thus motivational concepts are important in simple animals, because they act as models of processes in more complex animals.

5.3   Black box analysis

What, then, can be deduced about internal mechanisms by treating the animal as a black box and simply observing the relationships between stimuli (inputs) and behaviour (outputs)?

5.3.1   Stochastic threshold models

One simple theory is the threshold model shown in Fig. 5.3 which has been used to explain courtship in *Drosophila melanogaster* (Bastock & Manning 1955) and pecking preferences in chicks (Dawkins 1969). The idea is that a single random process, *V*, fluctuates with respect to certain thresholds. These are arranged such that at a low level of *V* no courtship is shown while at a high level of *V* all the components are seen; at intermediate levels some but not all components occur. The model excludes some classes of event; for example, if the thresholds are ordered as in Fig. 5.3, Licking must occur at the same time as Orientation and Wing Vibrating, and only Orientation can occur on its own. Any exceptions to these rules must lead to the rejection of the model.

Dawkins and Dawkins (1974) applied this idea to pecking preferences in chicks (Fig. 5.3) in a quantitative way. They showed that, by counting the numbers of pecks delivered to different stimuli (red, green and blue wheat grains) presented in different combinations, a coherent ordering could be obtained on the vertical axis, which means that a single random variable can account for the chick's choices. To explain the actual sequences of pecks it was necessary to postulate another random process, the 'go/no go' mechanism which determined when each peck occurred in relation to *V* (shown as the bottom line of Fig. 5.3). They were able to fit parameters describing these two processes to their

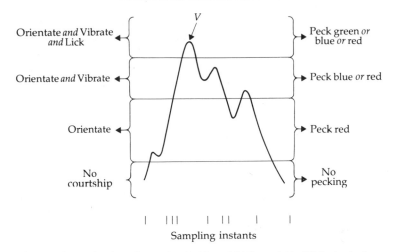

Fig. 5.3. Generalised stochastic threshold model. The variable (*V*) fluctuates in an unspecified way, interacting with thresholds corresponding to behavioural outputs. Outputs for *Drosophila* courtship (Bastock & Manning 1955) are on the left, and those for the chick pecking model (Dawkins & Dawkins 1974) on the right. Note that while the *Drosophila* outputs are superimposed literally, the pecking outputs are superimposed only in a probabilistic sense: the chick does not peck Red and Blue simultaneously but goes into a state of being equally likely to peck either. (From Dawkins & Dawkins 1974.)

data, and showed that the hypothesis that the 'go/no go' process was a relaxation oscillator, i.e. that the probability of a peck was low just after a peck but gradually increased, fitted the data well.

5.3.2   Statistical identification of internal processes

Although the stochastic threshold model of Dawkins and Dawkins (1974) gives a parsimonious predictive account of an animal's behaviour (which is all they claim for it) its relationship with real processes inside the chick is not very clear. Another type of approach is to look for statistical patterns in the behaviour of an animal and to try to relate them to underlying unobservable processes. The classic study of this type is the work of Wiepkema (1961, summarised by Hinde 1970) on the reproductive behaviour of the bitterling (*Rhodeus amarus*). Wiepkema used a statistical procedure known as factor analysis applied to the frequencies with which pairs of actions preceded and followed each other to find the underlying structure of the behaviour. He showed that a substantial part of the variation in the sequences he observed could be explained by only three underlying processes, which he tentatively identified as a sexual factor, an aggressive factor and a 'non-reproductive' factor. Although of great theoretical interest the technique has not been very widely used, largely because of technical difficulties in interpreting the results and because large numbers of observations are needed (Frey & Pimental 1978).

A development from Wiepkema's work is that of Heiligenberg (1974) on the cichlid fish *Haplochromis burtoni*. As well as using a somewhat different statistical approach from Wiepkema, Heiligenberg also considered the problem of long-term changes in the underlying processes, which Wiepkema's method could not handle but which are clearly likely to be involved in motivation. He showed that the most economical description of the long-term interactions between seven frequently observed activities contains four underlying processes (Fig. 5.4). However, this is only one of a set of possible models involving four long-term processes, all of which would give the same results. To decide which one is right it is necessary to do experiments. For example, if an experimental manipulation, such as applying a sensory stimulus or a hormone, could be found which caused a long-term increase in activities 2, 3, 4 and 5 but had no effect on 1, 6 and 7 then this could be

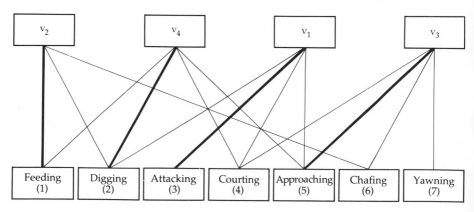

Fig. 5.4. The relationship between the per minute occurrence of seven commonly observed activities and four hypothetical slow processes v_1, v_2, v_3, v_4, in *Haplochromis burtoni*. The thickness of the lines indicates the strength of the dependence. (From Heiligenberg 1974.)

identified as a process analogous to v_1 (see Fig. 5.4). This restricts the possible models which fit all the data. An analogous experiment might allow the identification of v_2 and so on. Heiligenberg showed that a long-term process having the properties of v_1 is excited by the presentation of dummy fish having the black eyebar of a male *Haplochromis*.

5.3.3   Stochastic and deterministic theories

The analyses described in the last section all sought to give a probabilistic account of behaviour sequences. They could be described as statistical exercises to impose some order on the *a priori* uncertainty of behaviour (cf. Chapter 3) by fitting stochastic (i.e. probabilistic) models of the underlying processes. The main attraction of this approach is that it allows one to be agnostic about the nature of the internal mechanism where there is no direct information about it. Although the processes thus identified are flickering statistical ghosts compared with those seen in *Pleurobranchaea*, they are derived strictly from observations of behaviour and not from *a priori* notions about what the internal mechanism ought to be like. At the least, stochastic theories may provide an objective way to classify actions into activities in a natural scheme; at best they may tell us how many discrete internal mechanisms to look for and what some of their properties are, but they can never do more than predict the likelihood that a certain action will occur under specified conditions.

Most of the ideas dealt with in the remainder of this chapter carry the implicit assumption that behaviour is a deterministic process, that is to say that if enough was known about the mechanisms involved it would always be possible to specify precisely which activity an animal would perform next and with what intensity. This claim, which must always be more of an act of faith that a reasoned judgement, implies that stochastic theories merely cloak imperfect understanding in the disguise of 'random processes'. However, there are apparent examples of true randomness in biological systems. For example the segregation of unlinked Mendelian alleles seems to be as random as the toss of a coin. In some ecological models on the other hand deterministic processes may lead to chaotic behaviour under some conditions; such chaos is more apparent than real, but the point is that behaviour which

does not follow a regular pattern does not necessarily imply underlying randomness, any more than superficial order necessarily
implies underlying determinism. The only sound reason for prefer-
ring deterministic theories is that they encourage one to go on
asking questions about underlying causes where a stochastic
theory might encourage one to give up.

5.4   Unitary behaviour systems

Almost all discussions of motivation start from the premise that an
animal can be said to be doing one and only one thing at any instant
and that over a period of time it executes a sequence of different
actions with discrete switches from one to the next. A further
postulate that is usually made is that each of these mutually incompatible activities is controlled by a specific set of causal factors.
This is really just a restatement of the commitment to determinism,
that is to say that behaviour is caused by antecedent conditions and
that the same conditions always lead to the same action. Usually
the set of causal factors considered consists of components that can
be measured by an observer. Thus drinking depends on various
aspects of the animal's water balance and also on factors such as
the accessibility of water and cues indicating its presence to the
animal. Any particular causal factor may influence several activities; a high ambient temperature may cause the animal to seek out
shade, to show an increased interest in water and to reduce its food
intake. However, it is usual to exclude the possibility of causal
factors which have completely generalised effects on all behaviour,
such as 'arousal' or 'general drive' (see section 4.5). This position
underlies much current thinking about motivation, but is it in fact
correct? There are several points to think about.

5.4.1   Can animals do more than one thing at a time?

The answer to this question is partly a matter of definition (see
section 5.1). One can always classify behaviour in such a way
that only one action occurs at a time by definition, even though
the resulting classification may not correspond very well with
common sense. Fentress (1972) suggested that although animals
may only pay attention to one thing at a time, some activities can
become self-organising so that the animal can continue performing

them while attending to something else, rather as one can drive while attending to what is on the radio. He found that when grooming in a familiar environment mice responded to peripheral feedback from the face and whiskers, but that when they were put in an unfamiliar place their grooming became more stereotyped and was not affected by changes in feedback from the periphery. He concluded that they were effectively doing two things at once, grooming and visually exploring the new environment, but that the grooming was self-organising. However, examples such as this are rare and, in general, we can conclude that no great violence to the facts is caused by the assumption that animals do one thing at a time.

5.4.2   General and specific factors

It is now generally agreed that most activities are dependent on specific causal mechanisms. The hypothesis that there are general activating systems affecting all activities, such as Hull's general drive or arousal, has been effectively laid to rest, partly as a result of the failure to demonstrate such factors experimentally and partly because it is no longer felt that a driving force is needed to explain behaviour (see also section 4.5). Using engineering as an analogy, this is because motivation is now seen as a problem in the control rather than the powering of behaviour (McFarland 1966; Bolles 1975). However, there remain some empirical findings which do not fit very comfortably into the 'specific drive' mould.

Under some circumstances a mild but irrelevant stimulus may increase the rate at which an activity is performed. A moderate tail pinch may make a hungry rat start feeding sooner, though a strong pinch would block feeding and produce escape behaviour. It seems that the extent to which such irrelevant stimuli affect an activity may depend on the degree of attention being paid to the activity. Thus, in an animal that is hungry and feeding at a high rate, feeding is more resistant both to the disruptive and to the facilitatory effects of tail pinching than it is in an animal that is feeding at a lower rate (Fentress 1976).

5.4.3   Frustration

When an animal is thwarted, that is prevented from doing something, or when a reward is not given when the animal has been trained to expect it, a variety of effects on subsequent behaviour may be seen. In such circumstances a rat becomes agitated, apparently searching round for the anticipated reward, rearing and biting at bits of the apparatus. The long-term effect of omitting the reward is that the animal gives up the behaviour which led it to the reward (extinction) but in the short term other activities are often facilitated. Some activities increase in frequency compared with that during rewarded trials, or compared with a control situation in which the animal has no expectation of reward. The rate at which rats attack a satiated conspecific placed in the cage with them is increased during extinction compared with the baseline level during training (Thompson & Bloom 1966). Other activities increase in intensity; the classic demonstration of this 'frustration effect' is that of Amsel and Roussel (1952), extended and confirmed by Wagner (1959). They trained rats to run down a runway to a goal-box in which food was found and from which they could run down a second runway to another goal-box also containing food. A control group never experienced food in the first goal-box. The fundamental result is that when animals trained to expect food in the first goal-box find it missing they run faster along the second arm of the apparatus than they did when it was present, and faster than control animals who had never experienced food in the first goal-box.

There are a number of reports of rats drinking more than usual during food deprivation (Morrison 1968; Oatley & Tonge 1969) or when their daily food ration was not provided at the usual time (Panksepp et al. 1972). More remarkable, perhaps, is the phenomenon of schedule-induced polydipsia (Falk 1971), in which rats subjected to very intermittent schedules of food reward are found to drink vast quantities of water, which is voided soon afterwards as dilute urine. Although it was initially thought that any activity for which an opportunity was allowed might become schedule induced and so be performed in an excessive manner, and hence that intermittent schedules somehow produced generalised activation of all behaviour, such claims now seem to have been based on studies carried out with inadequate controls (Roper 1981). In rats schedule induction of drinking and of the gnawing of wood are the only effects produced reliably under properly controlled conditions.

These phenomena have been attributed to a generalised activation of behaviour, frustration, that results from thwarting. However, the results scarcely warrant this conclusion since the effects are clearly diverse. In most studies only one 'induced' behaviour has been measured, so claims that frustration is generalised in its effects are dubious. That there are some excitatory effects of thwarting seems clear however. It is probably relevant that non-reward seems to be aversive since rats will learn to escape from it or from cues which have been associated with it (Daly 1969). It is possible that some induced behaviours are ways of coping with aversive stimulation, though it is not clear how they help the animal to cope.

In summary, although thwarting of one activity may undoubtedly have facilitatory effects on some other activities, the effects are diverse and lend little support to the idea of generalised activating factors in behaviour.

5.5 Kinds of interaction

Interactions between activities may occur at one or more of three levels, distinguished by McFarland and Houston (1981) as the primary or peripheral level, the secondary or central level, and the level of the *behavioural final common path* (Fig. 5.5). The degree to which different behavioural subsystems interact varies from extremely closely related systems like feeding and drinking to

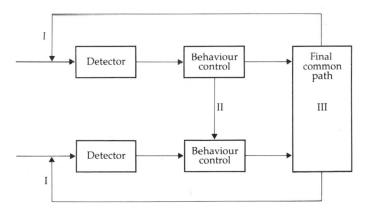

Fig. 5.5. Schematic representation of the levels at which interactions between activities can occur. I, Primary or peripheral level (e.g. feeding creates a water deficit). II, Secondary or central level (e.g. thirst inhibits feeding). III, Behavioural final common path, activities require incompatible use of the effector mechanisms.

those which scarcely seem to have any effect on one another such as grooming and exploration.

5.5.1   Interactions at the periphery

Although the ostensible function of an activity may be to correct some imbalance in the physiological state of the animal or to secure access to some resource, it will also have other consequences. A good example is supplied by the relationship between feeding and drinking, two activities which interact a great deal at the physiological level and which have been extensively studied (see Toates 1980a). The basis of the interaction is the fact that food intake creates a need for additional water intake in many animals through biochemical and osmotic processes. If protein is ingested, then the urea formed requires water to flush it away, and the ingestion of dry food pulls water into the gut osmotically thus creating a systemic water deficit (Oatley & Toates 1969). In a wide range of animals, feeding and drinking are closely associated in time (Kissilef 1969), although the association may be partly due to synchronised circadian rhythms in the two activities (Toates 1979a). The quantitative relationship between feeding and drinking depends both on the species and on such environmental variables as the water content of the food and whether or not free water is available. Water content of the food may be as high as 90% in lettuce or as low as 8% in laboratory chow. Gerbils need not drink free water when fed on fresh lettuce and fruit but must do so if fed on dry chow.

When rats are deprived of food, the amount of water they drink slowly declines, though there may be an initial increase in intake (Morrison 1968; Toates & Oatley 1972). The decline is what one would expect given a reduced need for water, but the results of food deprivation are not always so adaptive. In rabbits, food deprivation causes an increase in drinking due to loss of sodium which they are unable to retain in the kidneys. The extracellular sodium concentration falls so that an osmotic gradient occurs between the cells and the extracellular fluid, water is absorbed by the cells and the extracellular fluid volume decreases, provoking drinking. Since this creates a further osmotic gradient the process continues, and drinking cannot be satiated unless sodium is added to the water or food is provided (Cizek 1961, and see Toates 1980a).

5.5.2   Interactions within the nervous system

Many of the interactions between feeding and drinking do not seem to be fully accounted for by peripheral mechanisms. In rats the 1 ml of water which is typically consumed with 1 g of food is usually taken before the osmotic effect of the food could be operative (Oatley & Toates 1969). It appears that such anticipatory controls can be developed as a result of experience. Fitzsimons and Le Magnen (1969) shifted rats from a diet high in carbohydrate to a diet high in protein. The rats increased their total daily water intake during the first day after the switch in diets but at first the extra drinking occurred some time after feeding. After 3 days most of the extra water was taken in association with food; it appeared the rats were learning to anticipate the consequences of feeding for their water balance. However, this interpretation has not been checked by actually measuring the osmotic consequences on the blood of feeding (Rolls & Rolls 1982). In dogs with *ad libitum* access to moist food and water, drinking typically occurs 20 to 60 minutes after feeding and is preceded by changes in body fluids which would readily cause drinking if induced by direct manipulation; these animals cannot be said to show anticipatory drinking (Rolls *et al.* 1980), and in this case the interaction seems to be at level I (Fig. 5.5).

Since eating creates a need for water, it is not surprising that thirsty animals do not feed if kept on dry food, since this would make their situation even worse. In rats the reduction is quantitatively precise: the equivalent of 1 ml of thirst reduces food intake by 0.33 g, whether thirst is induced by deprivation (Toates & Oatley 1972) or by injection of hypertonic saline (Oatley & Toates 1973). The depression of feeding due to thirst appears to depend on inhibition within the central nervous system and not on competition between feeding and drinking for control of the effector mechanism. Angiotensin II is a hormone which is produced in the blood as a result of the release of renin from the kidney in response to a drop in blood pressure or a reduction in blood volume. It is a powerful dipsogen and when applied intra-cranially it not only induces avid drinking (Epstein *et al.* 1970) but also inhibits feeding (McFarland & Rolls 1972). The idea that the animal stops feeding simply because it is drinking all the time can be excluded by the fact that feeding ceases even when water is absent and there are no cues and no expectation of water in the experimental apparatus. The dipsogenic effect of intra-cranial angiotensin can be abolished by preloading the subject with water and this treatment also abolishes the inhibitory effect on feeding, implying that angiotensin inhibits feeding via its effect on the central mechanism controlling drinking. The conclusion is that the drinking control mechanism itself exercises inhibition over feeding during angiotensin-induced drinking (Rolls & McFarland 1973), but the role of angiotensin in normal deprivation-induced drinking is at best equivocal. Rolls and Wood (1977) showed that the removal of the renin–angiotensin system has no effect on deprivation-induced drinking and suggested that either it is an emergency back-up or else that the thirst mechanism contains redundancy, with cellular dehydration acting in parallel with the angiotensin system. It is not known whether thirst induced by cellular dehydration inhibits feeding in the same way as angiotensin-induced thirst.

Feeding and drinking are interesting systems to study because we have a fair understanding of the physiological mechanisms involved and because they provide illustrations of interactions at the peripheral level and of more subtle phenomena such as anticipatory controls which seem to implicate the central nervous system. Much of the rest of this chapter concerns motivational systems about whose physiology much less is known and whose interactions are ascribed to the behavioural final common path.

5.5.3   *The behavioural final common path*

The behavioural final common path represents the coming together of all behavioural tendencies to a common effector mechanism (McFarland & Sibly 1975; von Holst & von St Paul 1963). It can be thought of as the mechanism that decides which of the various incompatible activities for which the antecedent conditions exist will actually be expressed. When a foraging great tit (*Parus major*) is confronted by an intruder on his territory he immediately stops feeding and starts to threaten it. The physiological circumstances which led it to feed and the external stimuli relevant to feeding remain the same, but feeding is replaced by another activity. This type of interaction is often called behavioural inhibition, by analogy with inhibition at the level of neurons, on the grounds that some kind of physiological interruption must have occurred in the causal chain leading from stimuli to action. Hinde (1970) defines 'behavioural inhibition' thus

'Behavioural inhibition is . . . said to occur when the causal factors otherwise adequate for the elicitation of two (or more) types of behaviour are present, and one of them is reduced in strength because of the presence of causal factors for the other. Since it is usual for causal factors for more than one type of behaviour to be present, some degree of behavioural inhibition probably occurs all the time.'

According to this definition behavioural inhibition must play an important role in determining behaviour sequences. Since very little is known about the physiological nature of the decision process in the final common path we are again faced with the problem of discovering the properties of an unseen mechanism from the observation of its input–output relations. The difference from the work discussed in section 5.3 is that here we are examining deterministic models of the mechanism rather than looking for sets of statistical entities to explain the variability in behaviour.

5.6   Switches in behaviour

Since we are assuming that only one activity can be pursued at a time, it is useful to think of a ranking among activities such that the one with the highest rank is the one that has control of the behavioural final common path (cf. Atkinson & Birch 1970). The attribute of an activity used to rank it can be described as its 'strength of candidature' for control of the final common path; a more convenient term is the 'tendency' for the activity. This view of behaviour is deterministic in the sense that the activity shown depends in a systematic way on causal factors. Hinde (1970) used the term 'tendency' in a looser way, with increasing tendency meaning just increasing likelihood that the activity will be performed. Strictly speaking the term should probably only be used in relation to the elements of behaviour that are incompatible, i.e. actions (see page 136), but since the term is universally applied to activities in the literature I shall do the same.

There can be no direct observation of the tendencies for those activities which the animal is not currently performing, although they must be less than the tendency for the selected activity. The only way to observe them is to try to understand what happens at switches between activities, and to study the effects of manipulations of causal factors on the temporal organisation of the activities. When a switch in behaviour occurs it could be attributed largely to an increase in the level of causal factors or tendency for the second activity, which is then said to 'compete' successfully with the first activity for control of the effector mechanism; alternatively, the inhibition (in Hinde's sense, defined above) which prevented the second activity occurring before the switch could be removed by factors not relevant to the second activity, for example because of satiation or thwarting of the first activity. This is disinhibition, the essence of which is the idea that the start of an activity is somehow not solely dependent on its own causal factors (McFarland 1969). The term is used in this sense by many earlier workers such as van Iersel and Bol (1958), Rowell (1961) and Sevenster (1961).

One thing that is certain when an animal switches from activity 1 to activity 2 is that the ordering of tendencies for 1 and 2 has reversed; this general case is represented in Fig. 5.6. The problem is to discover what happened in detail in the region of the dashed lines in Fig. 5.6. The top row of Fig. 5.7 shows some of the logical possibilities. The different slopes (representing rates of change of tendency) for activities 1 and 2 will determine the effect that each has on the timing of the switch. The limiting case is where the tendency for one activity changes precipitately such that one can say that timing of the switch is independent of causal factors for the other (e.g. Fig. 5.7b). The decision about when such a condition arises depends on, amongst other things, the resolution used in examining the sequence, but a switch whose timing is indepen-

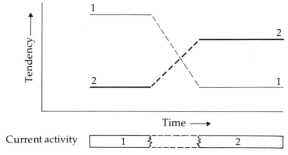

Fig. 5.6. A switch from activity 1 to activity 2 implies a re-ordering of the tendencies for 1 and 2 (see text).

dent of causal factors for the second of the activities involved corresponds to a switch involving disinhibition.

Unfortunately, a good deal of confusion exists over the labels used to describe the various kinds of switch which could occur during sequences of behaviour, in spite of two recent attempts to sort out the mess (Roper & Crossland 1982; Houston 1982). Houston (1982) suggested a new terminology to describe the relationship between the timing of a switch from activity 1 to activity 2 and the causal factors determining the tendency for 1 and 2 (Fig. 5.7). 'Dep$_1$' means that the timing of the switch depends on causal factors for 1, 'Indep$_1$' that it does not; there are corresponding terms for activity 2. He redefined the terms 'competition' and 'disinhibition' in his own way (Houston (a) in Fig. 5.7), but pointed out that one should really specify the effects of causal factors for both the activities involved in a switch (Fig. 5.7, bottom row).

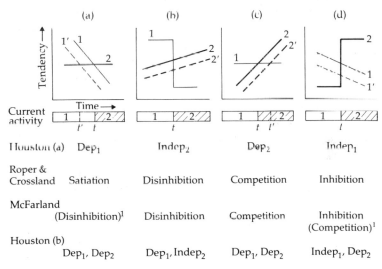

	(a)	(b)	(c)	(d)
Houston (a)	Dep$_1$	Indep$_2$	Dep$_2$	Indep$_1$
Roper & Crossland	Satiation	Disinhibition	Competition	Inhibition
McFarland	(Disinhibition)[1]	Disinhibition	Competition	Inhibition (Competition)[1]
Houston (b)	Dep$_1$, Dep$_2$	Dep$_1$, Indep$_2$	Dep$_1$, Dep$_2$	Indep$_1$, Dep$_2$

Fig. 5.7. A comparison of the terminology for behaviour switches used by Houston (1982), Roper and Crossland (1982) and McFarland (1969). The switch from activity 1 to activity 2 occurs when the tendency for 2 becomes the greater, at time *t*. If the tendency for one of the activities is altered by experimental manipulation the timing of the switch may be unaffected (b and d) or may be altered to *t'* (a and c). Below are shown four different sets of terms which have been applied to these situations. The use of the term disinhibition for (a) and competition for (d) ascribed to McFarland follows from some of his remarks, though he did not explicitly define the terms in this way.

The main lesson of Fig. 5.7 is that the terms 'disinhibition' and 'competition' need to be regarded with extreme suspicion wherever they are encountered. Although many switches in behaviour are, in Houston's terminology, 'Dep$_1$, Dep$_2$' there are quite a number of cases where a change is 'Dep$_1$, Indep$_2$', which all authorities agree is a sufficient condition for the label 'disinhibition', though it might not be a necessary condition. The slightly unexpected nature of such switches has led to the behaviour that follows the switch being labelled 'irrelevant', 'paradoxical' or 'out of context'; these are the characteristics of displacement activities.

5.6.1   Displacement activities

A displacement activity is an act which appears to be irrelevant or inappropriate to the context in which it occurs. Examples are the preening seen in chaffinches (*Fringilla coelebs*) when they wish to approach food but at the same time avoid a flashing light beside it (Rowell 1961), the ground pecking of male Burmese red junglefowl (*Gallus gallus*) during fights (Feekes 1972), or the nest fanning sometimes shown by male sticklebacks (*Gasterosteus aculeatus*) during courtship (Sevenster 1961). They are often seen when an animal is apparently uncertain what to do next because of contradictory stimuli, or when it is thwarted or prevented from carrying out the activity which would be expected to occur next.

The main criterion for recognising a displacement activity is its 'irrelevance', since in other respects it usually resembles 'normal' behaviour, although it may appear 'hurried, stereotyped and incomplete' compared to the same activity occurring in its 'normal' context (Bastock *et al.* 1953). In practice, however, alleged functional irrelevance often turns out to be illusory. Wilz (1970) showed that the shift from predominantly aggressive to predominantly sexual motivation which normally occurs during courtship in sticklebacks did not occur if the male was prevented from fanning the nest or creeping through it, both activities previously thought only relevant if there were eggs in the nest. Similarly, Feekes (1972) showed that 'displacement' ground pecking during agonistic behaviour between two male junglefowl affected the outcome. Where the opponents were equally matched the winner was usually the bird which was able to perform most ground pecking.

At least some so-called displacement activities therefore seem to play an important role in behaviour and can scarcely be referred to as occurring out of context, though in the case of the stickleback it remains a mystery why such an elaborate mechanism is needed for changing state. A more accurate definition of displacement activities might therefore be 'behaviour which at some time or another has been considered to be out of context'.

Displacement activities are certainly a diverse category from the causal point of view. Andrew (1956), for example, suggested that 'irrelevant' toilet behaviour in his buntings (*Emberiza* spp.) might be the result of discomfort caused by the autonomic effects of thwarting. What makes displacement activities interesting are the ways in which they are influenced by different causal factors. Rowell (1961) showed that the timing and duration of bouts of displacement preening in chaffinches were dependent on the factors affecting the conflict between approach and avoidance and not on manipulations relevant to preening, such as wetting the plumage or making the bill sticky. However, the nature of the behaviour within the preening bouts was influenced by causal factors relevant to preening. Similar results were obtained by van Iersel and Bol (1958) for the displacement preening of terns (*Sterna* spp.) during conflict between incubating their eggs and fleeing, and by Sevenster (1961) for displacement fanning during stickleback courtship. Since relevant causal factors influence the actual performance of displacement activities it is no surprise that they are causally diverse, but in respect of their timing they are *prima facie* cases of the transitions described by Houston (1982) as Dep$_1$, Indep$_2$ (see previous section) or by many authorities as disinhibition.

5.6.2 Disinhibition during normal behaviour sequences

Since displacement activities are apparently a diverse group of phenomena, there is no reason to suppose that the underlying mechanisms are peculiar to them. Rowell (1961) pointed out that the grooming seen in chaffinches on winter afternoons between the birds waking up from a rest and starting to feed seemed exactly comparable in its causation to that seen in approach–avoidance conflicts, except for the longer time-scale. To what extent is the most interesting feature of displacement activities, the apparent involvement of disinhibition, found in 'normal' behaviour sequences?

In a trivial sense the environment often causes disinhibition-like switches in behaviour (see section 4.4). If an animal is stalking a prey which takes flight then it will stop stalking and do something else, but the timing of 'something else' is obviously determined mainly by causal factors related to feeding. The more interesting case from the present point of view is where external conditions apparently do not change yet switches in behaviour occur without being obviously related to the internal state of the animal.

Persistence and stability

A simple explanation of behaviour sequences is that, for each activity, 'tendency', in the sense defined above, increases while the activity is not being performed and decreases while it is. Hence the behaviour currently in control of the final common path will eventually be ousted because of changes in its own tendency and in those for other competing activities. One difficulty with this view is the problem of dithering between activities, also known as the problem of persistence, which was pointed out by Bolles (1975) and by Atkinson and Birch (1970). In a simple homeostatic system when two tendencies become equal the animal ought to switch rapidly between them, since whenever behaviour A starts the tendency for it begins to fall and becomes less than that for activity B so that A stops and B starts, and so on until both tendencies reach zero. In reality animals do not dither between activities in this way; they tend to perform behaviour in discrete bouts. There are several ways in which such persistence could be explained. The simplest idea is that the negative feedback reducing the tendency for an activity as a result of performing it is delayed (for example in the drinking behaviour of Barbary doves, Streptopelia risoria (McFarland & McFarland 1968)). Alternatively, the tendency for the current activity may actually be increased initially by the momentum of performing the activity or by a positive feedback effect. There is some evidence that feeding tendency in mice increases at the start of a meal (Wiepkema 1968). Sexual arousal increases during the initial appetitive phases of sexual behaviour (Beach & Whalen 1959). In some cases behaviour consists of a series of decision points at which the animal can select a new behaviour, interspersed with periods during which it continues the current behaviour regardless of changes in internal state or external cues. This seems to be true of the sequence of actions seen during drinking in chicks (Dawkins & Dawkins 1974; see section 3.2.2).

An alternative way of looking at the problem is to consider it from a functional point of view. If an animal constantly switches from one activity to another it will spend most of its time switching and very little time doing anything else. This may not matter much if it is feeding and drinking from adjacent hoppers in a cage but might be serious if food and water were in different places, or if one of the activities required a minimum investment of time to achieve useful results as is often the case in courtship, for example in newts (Halliday & Sweatman 1976). That animals do take such considerations into account has been demonstrated by Larkin and McFarland (1978), who showed that Barbary doves pecking for food and water in a Skinner box made fewer switches during a session when the effort required to make the switch was increased by erecting a barrier between the keys.

The persistence of the current activity often results in the timing of the switches in a sequence being to some extent independent of the activity after the switch; i.e. they look like Dep_1, $Indep_2$ or disinhibition switches. The effect is that there is no clear relationship between the precise timing of the behaviour sequence and the internal state of the animal.

Time sharing

McFarland (1974) pointed out that an alternating sequence of two activities separated by switches of the form Dep_1, $Indep_2$ followed by $Indep_1$, Dep_2 (i.e. disinhibition followed by inhibition) would have the special property that one of the activities would have no control over the timing of the sequence; it would occur, as it were, by 'permission' of the other. This would mean that the two activities would have a difference in status with respect to their access to the behavioural final common path, the first activity being 'dominant' over the second. McFarland (1974) coined the term 'time sharing' to describe this relationship between activities, and proposed it as a model to explain displacement activities and other similar phenomena.

Two conditions must be satisfied empirically for time sharing in this special sense to be said to occur. It must be shown both that normal causal factors for the putatively subdominant activity affect it in the usual way, and that the timing and duration of the bouts of the putatively subdominant activity are not affected by its own causal factors but are affected by causal factors for the putatively dominant activity. Rowell (1961) showed that both conditions were satisfied by displacement grooming in chaffinches (see above). Cohen and McFarland (1979) showed qualitatively that under some circumstances both conditions are satisfied by displacement nest-related activities during stickleback courtship. They also showed that nest-related activities could be dominant during courtship, a departure from earlier accounts (Sevenster 1961; Wilz 1970). Lefebvre (1981) showed that the first condition was satisfied by the grooming seen when a cricket (*Teleogryllus oceanicus*) is exploring a novel environment, but his evidence for the second condition is less clear. Although the onset of grooming is a striking example of a Dep_1, $Indep_2$ switch, the duration of the bout may be increased by causal factors for grooming, and it is thus not terminated by an $Indep_1$, Dep_2 switch. Brown and McFarland (1979) showed that both conditions were satisfied by the feeding which occurs during courtship and copulation when a hungry male rat is placed with an oestrous female, but that no matter how hungry the male was made, feeding did not become the dominant activity over courtship. From the functional point of view this is not surprising, since feeding can normally be started or stopped at any stage without any great cost to the animal whereas there is no point in an animal starting a bout of intromissions if he does not proceed to ejaculation as the effects of intromissions do not accumulate in the same way as do those of food. If intromissions are not sufficiently close together in time he will never reach ejaculation (Beach & Whalen 1959; see section 6.5).

5.6.3   Criticisms of disinhibition

As McFarland (1966) pointed out, disinhibition is straightforward in principle but vague in detail. Used as a description of a behavioural mechanism it implies that an activity starts when inhibiting factors which previously held it in check are removed, leaving it to occur as if the inhibiting factors did not exist. There is often an unstated assumption that, since normal causal factors for displacement activities can be shown to influence them in the normal way, the causal factors relevant to the activity during which the displacement behaviour occurs do not have any effect. If this assumption was essential then facilitatory effects, such as the increased drinking which occurs during the extinction of feeding in the Skinner box, and the effects of frustration (section 5.4) would be contrary to the disinhibition hypothesis (Panksepp *et al.* 1972).

In fact the evidence has always been that the form of displacement activities was influenced by their context as well as by their normal causal factors (Rowell 1961; Wilz 1970), and the disinhibition hypothesis is certainly not invalidated by the existence of such effects. However, it does contain major flaws. For example a mathematical formulation of the model proposed by van Iersel and Bol (1958) would not produce displacement activities (McFarland 1966, extended and confirmed by Ludlow 1976, 1980). McFarland (1966) also pointed out that most of the ideas about disinhibition related to conflicts between two activities and that there was no *a priori* reason to suppose that thwarting of a single activity would lead to disinhibition. Nevertheless, displacement activities, to the extent that they could be said to exist as a discrete category, seemed to occur with similar properties in both contexts. Why then is the notion of disinhibition still influencing ethology? Its continued success seems to be mainly due to its usefulness in explaining 'paradoxical' switches in behaviour. If we could resolve the paradoxes we could discard the clearly unsatisfactory idea of disinhibition and all that goes with it.

5.7 The state space approach: an attempt at synthesis

There is clearly a need for a unifying framework in which the various kinds of interaction can be accommodated without necessarily having to class diverse phenomena as all being due to 'disinhibition', 'competition', 'frustration' or whatever. No doubt all the types of interaction described in the last few pages occur sometimes. The question is whether we can fit them into a single theoretical structure and begin to understand when and why one or other of them occurs. A candidate for such a framework is the *state space approach* proposed by Sibly and McFarland (1974) and further developed by the same authors (McFarland & Sibly 1975).

They start from the postulate that an animal always performs the action for which the tendency (in the sense defined in section 5.6) is greatest. An obvious way to represent such an arrangement is in a vector space whose axes represent the magnitude of each tendency in terms of some common currency. The state of all the tendencies is then represented as a point in this space of tendencies (often referred to as the candidate space as the behaviours are all candidates for expression); the space will be divided by switching lines into regions where one or other action has the greatest tendency. Tendencies depend on a variety of causal factors; thus tendency to feed might depend on food deprivation, the availability of food and the likelihood of predation. The obvious way to represent all these causal factors is in another vector state space, this time with axes representing all the causal factors. The statement that causal factors determine tendencies then becomes equivalent to saying that there is a mapping from causal factor state space to the space of tendencies such that for any specific state of the causal factors there exists a tendency state (which is just an ordering of the tendencies), so that it can be determined which action will be performed.

The relationship between tendency and causal factors is illustrated in Fig. 5.8 in which, for simplicity, only two-dimensional spaces are considered. The curves represented in causal factor space (on the left of the figure) join up sets of causal factor states which lead to the same tendency. Thus a particular feeding tendency could be caused by a low level of food deficit and the presence of a highly palatable food item, or by a high level of food deficit in the presence of less palatable food (see sections 4.3.1 and 6.3.3). For each activity the causal factor state maps to the one or more axes of the space tendency as shown in Fig. 5.8; combining all values for tendency gives the state (represented as a point in the space) determining which activity will be performed. As behaviour proceeds, its consequences alter the state of the causal factors; since each causal factor state maps into the space of tendencies, a trajectory will be traced out there. When this trajectory crosses a switching line the animal will switch to a new behaviour. Hence, if the causal factor state could be determined, and the mapping to the tendency space were known, the animal's sequence of behaviour could be determined. A more detailed account of these ideas is given by McFarland and Houston (1981).

One of the attractive features of this formulation is that terms like competition and disinhibition become a part of a general class of phenomena relating causal factors to tendency; there is no longer any need for sweeping generalisations about which type of

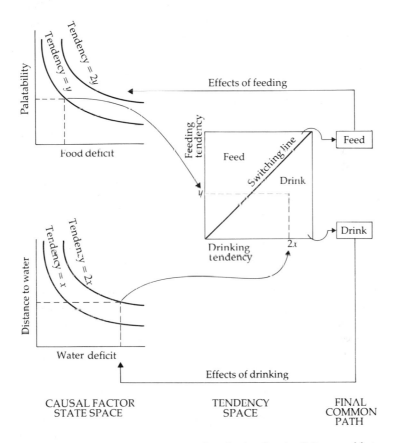

Fig. 5.8. The state space representation of motivation. For simplicity, causal factor state space is shown separated into the state variables relevant to feeding and drinking respectively; in fact a single state variable may influence the tendency for several activities. The curved lines in causal factor space join states leading to identical levels of tendency for a particular activity. In the case shown, causal factors relevant to feeding produce a feeding tendency of y, while causal factors relevant to drinking produce a drinking tendency of $2x$. These can be combined in the space of tendencies, from which it can be deduced which activity will be performed; in the case shown it will be drinking. The performance of an activity may influence the causal factors for itself or for other activities.

mechanism underlies behavioural sequences. It is not exactly a model of motivation in the sense of the term used in Chapter 6, neither is it a specific theory. It is a framework within which models can be set up and which shows that while hypothetical mechanisms like disinhibition, inhibition, time sharing and competition can be put forward as specific to particular systems, they are never the only way to resolve paradoxes in the sequences of an animal's activities. A single framework can incorporate physiological knowledge such as that described in Chapters 4 and 6 and in section 5.5 and interactions at the level of the final common path.

5.8   Functional considerations in motivation

An understanding of the sequence of activities pursued by an animal is bound to contain an essential element of functional explanation. Nothing in the laws of physics (or for that matter neurophysiology) could explain the calibration of hunger against sex, or aggression against thirst. In common with the study of structure and physiology, one way to investigate this functional aspect is to consider it as a design problem. Thus we investigate the engineering requirements for a given structure, and compare the actual animal with the design we would expect given the performance criteria. Analysis of structure using the 'Principle of Optimal Design' has given insight into the structure of many anatomical and physiological adaptations (see Alexander 1982 for examples). This principle states that, as a result of the evolutionary process, we expect animals to be well designed with respect to economy of materials and energy expenditure (Rashevsky 1960). The state space formulation provides the detailed analogy required to apply the concepts of optimal design to motivation, as it has been applied to structure, by allowing us to express the problem in terms of the design of complex control systems. The essence of the problem is the idea that there must be a trade-off between the advantage of performing one activity and the cost of not doing another (since only one activity can occur at once).

5.8.1   Optimality and behaviour sequences

One way to view the consequences of behaviour is to regard them as tracing a trajectory in the causal factor space following the sequence of states which occurs as a result of the behaviour of the animal. Such trajectories are predictable in principle though they may be hard to observe, and follow from the state space formulation outlined in section 5.7. Different calibrations of the relationship between causal factors and tendency would produce different behaviour sequences, so the functional analysis involves assigning some kind of value to the different trajectories theoretically possible and seeing whether the one the animal chooses is the best one or not. Since it is believed that the design process which gave rise to the actual calibration is evolution by natural selection, it follows that this value must represent evolutionary fitness in some way; the precise nature of this representation may be hard to specify since the fittest animal is likely to be the one that maximises its reproductive output over its whole life. In practice what is often done is to define a proximate goal, such as survival to the next breeding season or the next day, or to use some efficiency criterion such as 'maximise energy intake per unit of time spent feeding' and leave the exact relationship between this and lifetime fitness to be worked out later. The relationship which determines the 'cost' to the animal of being in a particular state for a period of time, or of performing an activity at a particular rate, is known as a 'cost function'. The nature of this relationship clearly has an important influence on which activity the animal should choose. Although cost functions are, in principle, measurable in the field there are great difficulties in practice. At the least, a colossal base of field knowledge would be needed and at present the feasibility of such an enterprise remains to be demonstrated.

The problem for the theoretician (which arguably evolution should have solved) is to choose the trajectory through causal factor state space which minimises the total costs incurred between the initial and final states—in mathematical terms to minimise the integral of cost, a mathematical function of state and rate of behaviour, over the whole trajectory. This is usually possible in principle, though it may require some sophisticated techniques (Houston 1980; McCleery 1978; McFarland & Houston 1981). The interest of all this from the point of view of motivation is that if the optimal trajectory is known it should be possible to deduce the decision rules, that is the relationship between causal factor state and the ensuing behaviour, which will give the best trajectory. This may enable us to understand the calibrations of one tendency against another.

McFarland and Houston (1981) list some recent applications of this dynamic optimisation theory; they fall into two categories. On the one hand are laboratory studies such as that of Sibly and McFarland (1976), who showed that the sequences of feeding and drinking bouts of Barbary doves recovering from food and water deprivation could be explained if cost is proportional to the square of the food or water deficit and to the square of the rate of behaviour. Such studies are truly functional analyses of motivation but suffer from a certain lack of plausibility. One may question how likely an animal is to show optimal behaviour under laboratory conditions, or indeed whether the precise sequence of eating and drinking bouts really matters to a Barbary dove. On the other hand are advances resulting from field data, such as the explanation of long-term changes in time allocated to feeding in weaver birds (Katz 1974) and the model of resource allocation to reproductives and workers in temperate wasps proposed by Macevicz and Oster (1976). In both these cases the predictions of the model are at a much more long-term level than the moment-to-moment switches between activities, so while they are more realistic from the point of view of measuring costs and benefits, they can scarcely be classed as motivational theories.

5.8.2   Retrospect

Although the idea of using the state space representation of motivation in conjunction with optimal control theory to unravel functional aspects of behaviour sequences is deductively sound, no thoroughly worked out example has yet been produced. It remains for the present a program of research rather than a synthesis of established knowledge. There are two aspects of the program which can be criticised: the applicability of optimisation theory and the plausibility of the state space model of motivation.

Applicability of optimality theory

The use of the term 'optimal' has generated a certain amount of heat (Gould & Lewontin 1979; Oster & Wilson 1978), mainly because it appears that some claim is being made about the perfection of animals or that a test is being made of whether animals are optimal in some global sense; this criticism applies to studies of morphological adaptation as much as to behaviour. In fact nothing of the kind is being attempted. What is under test is a specific hypothesis about how an adaptation works given certain constraints which may themselves be the result of natural selection or simply historical accidents (Maynard Smith 1978). However, even when this is accepted, some problems remain. Could the design process track small, rapid environmental changes, or would there be a constant lag behind the true optimum? The answer to this question must involve knowledge of rates of evolution and hence requires the specification of both selection pressures and the genetic mechanisms involved; the latter are never specified in detail. How can the goodness of fit between the actual design and the 'true' optimal design be evaluated? The statistical means of assessing this goodness of fit are not well defined and the problem is compounded by the fact that deterministic theories always produce deterministic predictions but real behaviour sequences always contain some unexplained variability. Even when a convincing fit to the predictions is found, the implications are not very clear since in most cases alternative models which make similar predictions are not even formulated, far less eliminated. The alternative result, where the predictions do not fit the data, is equally fraught with problems. The animal may fail to oblige because it uses rules of thumb which are not adapted to the test situation, or more generally because the cost functions or performance criteria have been incorrectly specified, or because important causal factors have been missed.

Against these difficulties must be set the undoubted success of optimality analyses in understanding how behavioural choices are adapted to their circumstances; this is particularly true in studies of diet selection and foraging to maximise net rate of energy uptake (Krebs & McCleery 1983; Krebs *et al.* 1983). It seems that many animals get near to optimum solutions, even where quite short-term changes in the environment occur, and they do not necessarily follow rigidly defined policies. Thus, although there are some problems in the use of optimality, it is proving a useful method for understanding functional aspects of motivation.

Adequacy of the state space model

All motivational theories incorporate the notion of calibration of different motivational systems against each other (Toates 1981). An animal performs the activity with the biggest deficit or the biggest drive, or it performs the most advantageous activity, the one with the biggest incentive associated with it or that which will yield the greatest reinforcement (Bolles 1975). Such theories agree that there exist comparable tendency variables, and they only differ in their account of what determines tendency.

The implication of this 'specific tendency' model is that tendencies are more or less continuously related to causal factors, so that as the causal factor state changes so tendencies for various activities fluctuate, with the largest at any time controlling the effector mechanism. It has already been pointed out that, taken literally, such a model is likely to dither eternally between activities unless specific mechanisms are introduced to prevent it, but here I want to consider the whole notion of instantaneous control in behaviour. Studies of causation using various kinds of control theory are often rather successful at predicting broad features of behaviour (e.g. at the level of 'rate per hour'), but deterministic theories of the fine structure of behaviour are strikingly unsuccessful at predicting actual sequences of activities, hence the proliferation of concepts like 'disinhibition' and 'time sharing'. This suggests that the idea that animals constantly evaluate tendencies and pick the biggest is at best an over-simplification. Halliday (1980), for example, suggested that rather than thinking of a sparrow feeding in a dangerous place as constantly measuring off tendency to feed against tendency to look up, viewing these as separate systems competing for expression, it would be more realistic to imagine it scheduling feeding and looking up so as to be able to feed while concurrently maintaining adequate vigilance. Hence the actual sequence is the product of a composite system with two goals rather than two independent systems in competition. This may seem a trivial distinction at first sight, but the example can be taken further. Suppose the sparrow is vulnerable to a predator like a cat which stalks its prey, waits until the prey looks the other way and then attacks with a final unconcealed rush. If the sparrow looks up during the attack it stands a good chance of escaping, so the best thing to do is look up at regular intervals such that the gap between scans is too short for an attack. However, it may be impossible to achieve this level of vigilance and still get any feeding done, in which case the best policy is to look up randomly so the predator cannot predict the probability of being spotted on the basis of the previous scan. Regular looking up would be disastrous since the cat could always fit in its attack after a scan. Here the trade-off between activities occurs not at the level of instantaneous decisions but at that of patterning over periods of minutes or hours. The actual behaviour sequence is a solution to the scheduling problem set by the medium-term requirements of the animal. No straightforward competing-tendency model could account for such a sequence.

Of course, it remains axiomatic that the animal does whatever has the highest tendency. It is the way in which tendency relates to causal factors that should be the focus of efforts to understand behaviour sequences. The criticism is therefore not of the state space model as such, but of the way in which it has previously been applied. It is a framework sufficiently flexible to incorporate sophisticated scheduling rules as additional causal factors but only at the cost of making 'tendencies' depend on causal factors in a rather complex way. While this conserves the definition of tendency, the term thereby loses some of its heuristic value. It remains to be seen whether the state space approach is another enormity (see page 134), but at present it seems to be the best way forward to a proper understanding of motivation.

5.9   Selected reading

The treatment of motivation by Hinde (1970) remains extremely authoritative (brought up to date in a less comprehensive form by Hinde 1982). Toates (1980a) combines a broad coverage of relevant physiological psychology with good accounts of ethological ideas in a framework of 'systems thinking'. McFarland and Houston (1981) provide a definitive treatment of the state space approach, and of optimality models. The latter topic is dealt with at a less technical level by McCleery (1978) and by Houston (1980). An excellent introductory account of optimality theory applied to several areas of zoology is given by Alexander (1982). The book edited by Toates and Halliday (1980) contains a number of useful papers, notably those by Roper on induced activities and by Ludlow on models of decision making. Houston (1982) provides a critical appraisal of the literature on time sharing.

CHAPTER 6
MODELS OF
MOTIVATION

FREDERICK M. TOATES

6.1 Introduction

With different authors, the word 'model' takes on various meanings or shades of meaning. It is beyond my brief to go far into the philosophy associated with modelling, but a few words in this direction are needed. Emphasis will be upon models as *theories* or *formal statements* of how systems are believed to work. In this sense, the meaning has something in common with, say, a 'model' railway. In its performance the model railway is not, of course, identical to the real railway system. However, in their mode of operation these two systems have important features in common. From the model, one could predict operating characteristics of the real system, such as the tendency for two high-speed trains to scrape each other on taking a sharp bend. 'Model' and 'theory' are, in some respects, interchangeable terms. Ideally, one can make predictions from a theory and test them against the real system. This is indeed the case for theories that are built as models.

Models are translations from empirical observations to a particular formal language. In the study of animal behaviour, they are usually made up of either (1) symbols and equations or (2) boxes containing mathematical terms and interconnected in various ways. In both these cases the models are written on paper. However, a model can be constructed from electrical or hydraulic components; the actual embodiment is not important. The only major considerations are whether the model's components exhibit the characteristics of components in the real system, and whether the behaviour of the model can suggest anything about the behaviour of the real system. Take, for example, the hydraulic model of motivation developed by Lorenz (1950) and described in Chapter 4 (Fig. 4.3). If one were to construct this model from valves and water, then an unambiguous account of how the model behaves would be obtained. Test stimuli could be applied, and the responses noted. In practice, ethologists usually discuss Lorenz's model in terms of their knowledge of how such a system would behave, were it to be constructed. Then they relate this to motivational processes. These days, a common way of modelling is to build assumptions into a computer program and get the computer to predict the outcome of the assumptions. This technique will be described later.

An important aspect of a certain kind of modelling is brought into clear relief by Lorenz's contribution. In his model, he is not suggesting that, in reality, when animals are unable to perform a class of behaviour they accumulate fluid. Nor is he suggesting that, when consummatory behaviour occurs, fluid is released through a valve. What he *is* saying is that the processes underlying behaviour have important characteristics *in common with* such a system. It is an 'as if' model.

A kind of model rather different to that of Lorenz is where, on the basis of independent experimental evidence, we have established the properties of individual components of a system. In this case, we may need a model in order to give us an account of the way in which such components *interact* in the total system to produce behaviour. In other cases, we may have only a vague knowledge of the characteristics of the individual components or know the identity of only some of them. Still, as with Lorenz's model, it might be useful to speculate as to what kind of components the system might employ.

Sometimes models may mislead us, and it is as well to be aware of this. A model may work, in that it yields correct predictions, but still be based upon false premises, in that the characteristics of its components have nothing in common with those of the real system. An example might be that a human voice and a gramophone playing a record sound very similar, but clearly the subsystems involved have little in common. Another danger may arise even if the components have realistic characteristics. As we stated earlier, for some purposes the physical embodiment of a model need not be important. However, occasionally, an 'as if' model is believed literally. In the extreme, this might, for example, lead us to search for a real motivational fluid chamber and valves in the head of an animal.

Most models should not be believed too seriously. Rather, they should be challenged regularly to see whether a better one can be devised. That is, a new model may fit a wider range of data. In the study of behaviour, models at best capture only 'features' or 'aspects' of how complex systems work, and a healthy scepticism is advisable.

In the study of motivation, models serve various purposes. As one example a model might be devised to show how a particular motivation (e.g. the tendency to drink) arises from the organism's physiology. Other models attempt to show how different motivations compete for expression in behaviour, and pay little or no attention to how any particular motivation arises. In this chapter, I shall give examples of various kinds of model. Each of these was constructed with a particular problem in mind and with certain implicit assumptions. From the spectrum of models described, the reader should gain an idea of the value and applicability of the modelling approach. First, however, I shall introduce some of the terminology with which it is necessary to be familiar.

6.2   Terminology

6.2.1   *Block diagrams*

Models deal with a flow of *information* (Toates 1975). Such information often denotes physical causality: events at point a influence events at point b in the manner shown by the model. Information shown in a model may simply denote a relationship. This is best explained by examples. Figure 6.1a shows an arrow, the input, entering a box and another arrow, the output, leaving the box. The box itself contains an operator. What this means is that a particular signal, the input, is operated upon and transformed to give another signal, the output. The transformation may be a complex one in which we must guess *how* the operator achieves the conversion. For example, a term such as 'memory' may, rather vaguely, summarise a transformation. Alternatively, the transformation may be very simple and unambiguous. For example, if we wish to convert a distance in inches to one in centimetres, we simply apply the transformation: centimetres =

inches×2.54. In Fig. 6.1b we show this operation. A term (in this case 2.54) that multiplies an input to give an output is sometimes called the *gain* of a component. Suppose that, in this language of modelling, we wish to represent the relationship between the volume of water in a container and the flow into the container. Volume depends upon flow, so the causal link is that we transform flow in order to give volume, assuming that the container is empty at the time that we start our observations. This transformation is shown in Fig. 6.1c by the ∫ sign. This sign means that volume is given by the integral of flow with respect to time. For the reader not familiar with mathematics, Fig. 6.1d explains the meaning of this

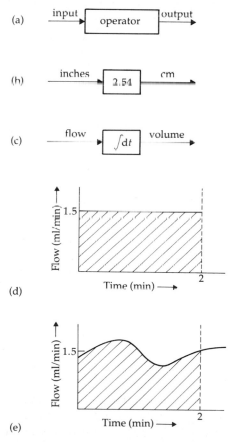

Fig. 6.1. The notation and some results of model building. For explanation, see text.

term. If the flow is 1.5 ml/min for 2 min then the volume will be 3 ml. This is given by the shaded area, and we have performed a simple case of integration. Consider though the fluctuating flow shown in Fig. 6.1e. In this case, one cannot perform a simple multiplication to obtain the integral. However, as before, the integral is given by the shaded area. So one need do no more than imagine a mathematical device that works out the shaded area, and this would calculate the integral. The integration sign is a

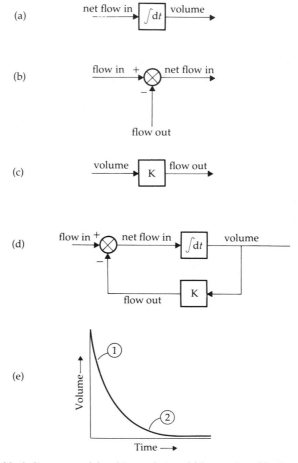

Fig. 6.2. A block diagram model and its prediction. (a) Integration, (b) subtraction, (c) multiplication by a constant, (d) model of a leaky container, (e) result predicted for the volume of fluid in the container.

mathematical operation that is synonymous with 'the area under the graph after a period of time has passed'.

Now imagine that the container has a hole in the bottom, through which water leaks. At the time of starting the observations the container is empty, then a tap is turned on and water enters. The level rises and water escapes through the outlet. Can we construct a model that represents this system, showing the variables and the relationship between them? Let us start by looking at the component operations. Flow *into* the container is no longer the only factor determining the volume; flow *out* is also a factor. So, in order to obtain volume, we need to replace 'flow in' by 'net flow in', and integrate this (Fig. 6.2a). Net flow in is given by flow in minus flow out (Fig. 6.2b). The summing junction (a circle with a cross in it) and its associated ı and signs indicate this. Presumably, flow out depends upon the pressure of fluid; as the container fills, flow out increases. If flow out is proportional to volume then

$$\text{flow out} = \text{volume} \times K \qquad (1)$$

where K is a constant. We can incorporate this assumption as shown in Fig. 6.2c. Figure 6.2d shows the whole model put together. Now what we have done is to present in clear and unambiguous terms our model of how the system works. Our assumptions can be seen at a glance from Fig. 6.2d. Suppose that we wanted to predict the behaviour of our model in response to a particular stimulus, for instance a sudden load of water. We could easily work this out mathematically, or get a computer to do it for us. Figure 6.2e shows the change in volume predicted by the model in this event. The rate at which volume falls is, of course, the flow rate out of the container. Flow out is proportional to volume (equation 1). Thus, when volume is high (e.g. at 1), flow out is high, whereas later on (e.g. at 2) both are low. A curve of the kind shown in Fig. 6.2e is known as an *exponential decay*.

6.2.2 Feedback

A term that you will often meet in a discussion of models is feedback. The meaning of this term is illustrated in Fig. 6.3. A lever is free to swing around a pivot. The lever would naturally occupy the position shown in Fig. 6.3a. Suppose, by a brief tap, the lever is displaced to the position shown in Fig. 6.3b. It has been displaced by an angle θ, and, as a result, a force F is introduced. This force tends to return the lever to its former position. The general point is that a displacement tends to introduce a force such that the displacement is either eliminated or reduced. This is an example of a particular class of feedback system, showing *negative* feedback. The word 'negative' refers to the fact that disturbances tend to eliminate themselves. Sweating in response to a rise in body temperature is an example of negative feedback. The rise causes sweating, and sweating tends *to counter* the rise. By contrast, Fig. 6.3c illustrates positive feedback. The lever in Fig. 6.3c is balanced precariously on its end. Any tiny disturbance θ creates a force that increases the size of θ. Such positive feedback is short-lived. The lever soon falls to the stable position, hanging down, a situation in which negative feedback is applicable.

6.2.3   Feedforward

Negative feedback control is action taken in response to a disturbance so as to correct the disturbance. Feedforward is action taken in *anticipation* of a disturbance and, if successful, it *prevents* the disturbance from occurring. For example, in response to a cooling of the body we might put on extra clothes. That would be feedback. However, we might put on the extra clothes when told

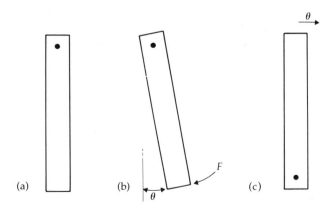

Fig. 6.3. Negative feedback (a,b) and positive feedback (c) illustrated by a rod pivoted at one end about a nail.

that the weather is about to get colder, and that would be feedforward. Of course, both processes can be present at the same time, as they are in behavioural temperature regulation in humans, and one or the other may be more evident according to circumstances.

6.3 Feeding

Feeding provides some rather nice examples of where a modelling approach has been used and where concepts associated with modelling are at the centre of discussion.

Feeding is normally under the joint control of factors internal (e.g. body energy detected at some site) and external (e.g. cues from food) to the organism. Some models have been designed to show how these two types of factor interact to determine ingestion. A few examples will illustrate the way in which modelling has been helpful

6.3.1 Feeding in flies

In a stimulating and entertaining review, Bolles (1980) considers the most simple model that could represent the essence of the feeding system of the flies studied by Dethier (1976; see Fig. 6.4). A gustatory stimulus g excites feeding, via operator K_1. The product $g \times K_1$ gives the strength of excitation. Consider what happens when the fly eats. Food intake is integrated to give the gut content (for simplicity ignore the emptying of the gut); this is multiplied by K_2 to give the strength of inhibition on feeding. This loop exhibits negative feedback: feeding has effects that terminate feeding. The summing junction compares inhibition with the strength of excitation. If excitation is greater than inhibition, the fly ingests. Feeding continues until inhibition terminates ingestion. Simple as it is, Bolles argues that, on the basis of Dethier's experimental results, Fig. 6.4 nonetheless represents rather well the essence of the fly's feeding system. He notes that if one cuts the appropriate nerve, an operation indicated by the dotted line, then one eliminates inhibition on feeding and the fly will literally burst from overeating. Normally, with abundant food, the fly functions rather well, thanks to the inhibitory loop. It does less well at times of food shortage. Figure 6.4b shows a range of food availability and its relevance to survival. Provided that availability does not fall below A_1, the fly can survive, since intake level I_1 is sufficient. Intake rises no higher than I_2, even when availability exceeds A_2. This is because of the inhibitory link. Were this link not present, the fly would enter the range shown by the broken line and would burst.

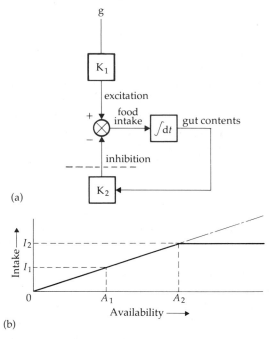

(a)

(b)

Fig. 6.4. Feeding in flies. (a) Model of the system, (b) intake of food as a function of food availability. (Modified from Bolles 1980.)

This model demonstrates how, on the basis of a few simple assumptions, one can make predictions concerning the behaviour of an animal. It may be that the system will turn out to be much more complex than this, but Bolles has shown that in principle a few simple assumptions are sufficient to account for the results.

6.3.2   The Wirtshafter–Davis model

This model, constructed primarily with rats in mind though of broader application, is shown in Fig. 6.5 (Wirtshafter & Davis 1977). It is obviously a simplification, but nonetheless makes some

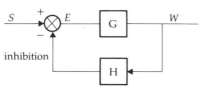

Fig. 6.5. Model of feeding proposed by Wirtshafter and Davis (1977).

interesting points. It is based upon the assumption that a tendency to feed is aroused by the sensory properties of food. The strength of this signal is shown by S. A measure of the body energy state, such as the size of the fat stores, is represented by the variable W. This exerts, via term H, an inhibitory influence on feeding tendency; the larger the energy store, the stronger the inhibition. Whether or not the animal eats depends upon the difference between excitation and inhibition (E). The parameter G (which is in fact just shorthand for an extremely complex set of many terms) relates E to the size of the energy store (W). Suppose the animal has not eaten for a long period of time, and then finds food. W is low, hence $S - WH$ is high and so E is high. Thus, over a period of time, the animal eats a substantial amount. Conversely, if the animal has been force-fed, then W is unnaturally high. In the presence of food, E would be negative and the animal would refrain from eating. On the basis of laboratory studies, it is clear that acceptable and nutritionally valuable diets vary in their palatability. Presumably, in the wild, the foods that animals such as rats are prepared to ingest show a range of palatabilities. In the model, the magnitude of S represents how palatable the diet is. Suppose a rat is existing on a diet of medium palatability and then finds a diet of high palatability. Daily calorie intake will rise, and so will W. But as W rises so does the strength of inhibition. Thus, by this negative feedback effect, a limit on intake is imposed. Conversely, if only an unpalatable diet is available, the rat's intake falls, as will the level of its energy store. But, as this happens, the strength of the negative feedback declines, increasing E, and setting a limit on how far intake falls.

This model claims only to be relevant to an aspect of long-term regulation, and does not pretend to say anything about onset of individual meals. But it shows how regulatory phenomena (setting upper and lower limits on calorie intake) can *emerge* from the properties of a few components put in combination.

6.3.3   The McFarland–Sibly model

Some models (e.g. that of Bolles) show how component physiological operations, albeit simplified ones, can explain aspects of
motivation. The model discussed in this section (McFarland &
Sibly 1975) simply asks how two easily manipulated factors may
combine to give a motivational tendency. These factors are (1) the
quality and availability of food and (2) the hours of food deprivation. Although devised for feeding, the model, which is based
upon empirical evidence, may with suitable modification be
broadly applicable to other motivational systems.

This model was discussed in section 5.7 and is illustrated in Fig.
6.6. The strength of the tendency to feed is given by hours of food
deprivation times incentive. The term 'incentive' represents the
power of the external stimulus, food, to arouse ingestion. A high
incentive value might arise from highly palatable food (as in the
Wirtshafter–Davis model). It might also arise from a readily available food (e.g. one in close proximity) as opposed to food the
acquisition of which would require energy and time. McFarland
and Sibly use the term *cue strength* instead of incentive to cover the
power of the external stimulus as a factor in determining motiva-

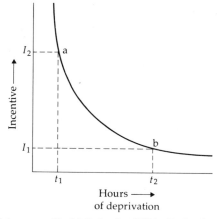

Fig. 6.6. The model proposed by McFarland and Sibly. Each point on the curve has
an equal strength of motivational tendency. For example, point a is obtained from a
combination of high incentive I_2 and a short period of deprivation t_1. Point b gives
the same motivational strength but is obtained from a low incentive I_1, and a long
period of deprivation t_2. Another curve would be needed to represent a
combination having a higher motivational tendency.

tional tendency. Note, in Fig. 6.6, that the same tendency can arise from various combinations of deficit and incentive (section 4.3.1).

This model can be used to give an example of *positive* feedback. Suppose an animal is in an ambivalent position, thirsty and hungry, mid-way between food and water. Now, suppose it moves towards the food. This increases the cue strength of food which, according to this model, increases the strength of the feeding tendency. This moves the animal even nearer the food. In this way stability is introduced into the system; the animal does not dither between feeding and drinking, achieving no substantial intake of either. If motivational strength depended solely upon internal factors, the animal would be expected to dither (section 5.6.2).

6.3.4 Physiological models of feeding

Various attempts have been made to produce computer simulations that are based upon detailed physiological components and which predict the occurrence and size of individual meals (Booth 1978). However, we have only a vague idea of what is the internal energy state that is detected and that determines meal onset. Various suggestions have been made, ranging from the ingenious to the bizarre. Two serious possibilities are that feeding is excited when blood glucose level is low (Le Magnen 1981) or when the rate of supply of energy at the liver falls below a threshold (Booth & Toates 1974). Figure 6.7 shows these proposals, and they are described in the legend. Both involve a clear negative feedback aspect. Food intake ultimately cancels the instigating stimulus, be it low energy supply rate or low blood glucose level. However, it takes time for ingested nutrients to reach the blood. Can the instigating stimulus be eliminated sufficiently quickly for this to account for the termination of a meal? This seems unlikely because many animals eat quickly and therefore must stop eating in *antici-pation* of the arrival of nutrients rather than in response to their arrival. For instance, in Fig. 6.7a an animal will eat when energy supply rate falls below (to the left of) e_1. If an animal is at e_x and finds food it is quite likely to stop eating before the rate reaches e_1. Why is this?

In, say, the rat, over a period of time learning appears to play an important role in terminating meals. An animal stops eating when

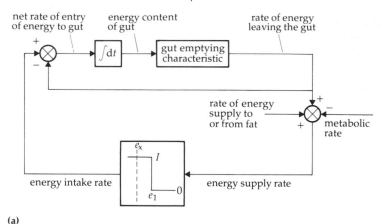

net rate of entry of energy to gut

energy content of gut

rate of energy leaving the gut

rate of energy supply to or from fat

metabolic rate

energy intake rate

energy supply rate

(a)

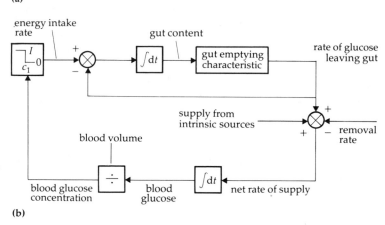

energy intake rate

gut content

rate of glucose leaving gut

supply from intrinsic sources

removal rate

blood volume

blood glucose concentration

blood glucose

net rate of supply

(b)

Fig. 6.7. Models of feeding. (**a**) Energy supply model. Start at the top left, with 'net rate of entry of energy to gut'. This is integrated to give 'energy content of gut'; the box marked 'gut emptying characteristic' relates this energy content to 'rate of energy leaving the gut'. This is the rate at which energy arrives at the liver. Metabolic rate is the rate at which the cells of the body burn fuel. Some energy is converted to and from fat. The net result of all this gives 'energy supply rate', which is high when a large amount of fuel is arriving from the gut or when intrinsic fat deposits are being broken down. When this rate falls below e_1 the animal eats, i.e. energy intake rate goes from 0 to I. This then forms a positive input to the gut integrator. (**b**) Model in which feeding depends upon blood glucose level. Blood glucose is the integral over time of all the inputs, i.e. rate of arrival from the gut, supply from intrinsic sources (e.g. breakdown of stored fuel), minus the rate at which glucose is taken from the blood (e.g. by the cells). Dividing blood glucose by blood volume gives blood glucose concentration. When this falls below c_1, intake rate shifts from 0 to I.

it has taken in x grams (at time t_1) because in the past when it has done this the arrival of a substantial amount of energy has been detected centrally at a later time (t_2) (Booth 1980; Toates 1980a; section 4.6.1). This is feedforward. Information on the arrival of food at the mouth and stomach feeds forward to switch off eating. At time t_1 some energy will have been absorbed, and meal termination appears to be the result of a combination of oral and other factors (Booth 1980). Metaphorically speaking, the animal says to itself 'stop now and be patient since a substantial amount of energy will arrive shortly'. Without feedfoward the animal would overeat.

Suppose the animal finds itself eating a new diet of a lower caloric density than that it ate before. The previously learnt strategy will now, to some extent, be inadequate. If it stops at time t_1, then at t_2 less energy will be detected centrally than on the richer diet. In practice, this is the cue for readjustment of the parameters; the animal must eat for longer than t_1. The strength of oral factors in leading to satiety needs to be lowered. Rats are very good at adjusting meal size in the light of such experience (Booth 1980). A similar effect of experience, leading to the avoidance of noxious foods, is described by Roper (1983).

6.4 A physiological model of drinking

In this section, we shall consider a model of the physiological bases of drinking (Toates & Oatley 1970). To a considerable extent, the model is built with components that can be characterised from independent physiological evidence. The aim here is not to provide a detailed discussion of the entire model, but to describe some of its subsystems.

For some (but not all) purposes, the body fluids may be considered to be divided into two fluid compartments—intracellular (or cellular) and extracellular. The intracellular compartment (ICF) includes the sum of all the fluid in the cells of the body, whereas the extracellular compartment (ECF) consists of all of the fluid outside the cells (including that in the blood). In response to a disturbance in the concentration of ions across the membrane around the cells, water is able to pass between the two compartments. It is reasonably accurate just to discuss the two principal cations, sodium and potassium, and to consider sodium to be confined to the extracellular compartment and potassium to be confined to the cellular compartment (Fig. 6.8). We divide the amount of sodium by the extracellular fluid volume to get extracellular sodium concentration, and the amount of potassium by the cellular fluid volume to get cellular potassium concentration. The constants K_1 and K_2 are simply scaling factors, so that we arrive at a steady state equilibrium after inserting the values of the ionic and fluid quantities that normally prevail. By subtracting one term from the other, we have the difference in concentration across the membrane, that is to say, the force tending to move water across the membrane. This is zero under equilibrium conditions. Following a disturbance, water may move across the membrane in either direction, depending on which concentration is higher. Assume that the flow of water is proportional to the concentration gradient. Parameter K_3 expresses this: gradient$\times K_3$ gives the flow. In turn, flow feeds back to change the volume of the extracellular

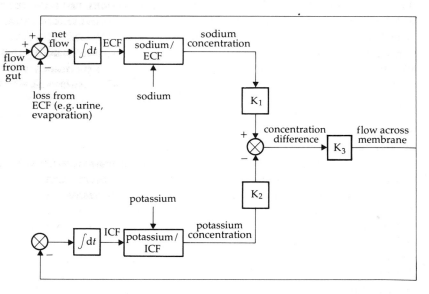

Fig. 6.8. Model of the exchange of fluid between extracellular (ECF) and intracellular (ICF) compartments. Note that the net flow into the ECF depends upon the flow from the gut, loss from the body and gain from the ICF. If the concentration difference is negative, then this causes a flow from ECF to ICF. For example, consider a sudden flow of water from the gut. ECF is elevated, sodium/ ECF decreased and concentration difference is negative. Therefore, the term 'flow across the membrane' takes a negative value. The minus sign at the input to the ICF integrator further negates this which means a *positive* input to ICF.

and cellular compartments, adding to one the amount that it subtracts from the other. This diagram now enables us to model water movement following, for example, an injection of concentrated sodium chloride into the extracellular compartment. Another example would be where an animal becomes depleted in sodium, so that the quotient Na/ECF is lowered. This would result in a net migration of water from the extracellular to the cellular compartment. Total extracellular volume is therefore depleted and the cells become over-hydrated.

Consider now another subsystem of the complete thirst/fluid model, the part relating directly to motivation (Fig. 6.9). This shows the arousal of drinking by particular body fluid states. In Box 1, a reduction in extracellular volume below some normal level has an excitatory effect on drinking. For example, volume x_1 is associated with a magnitude of excitation given by y_1. Similarly, in Box 2, a reduction in cellular volume to below normal has an excitatory effect on drinking (just as hypertonic saline injections arouse drinking). Either stimulus can arouse 'drinking motivation'. By the positive signs associated with the arrow tips at the summing junction, the two influences are shown as adding to give a total excitatory effect, as happens following a period of water deprivation (Fitzsimons & Oatley 1968). The box marked 'drinking mechanism' relates this excitation, if it is sufficient, to the act of ingestion of water. This subsystem implies *negative* feedback: a displacement causes action (drinking) that corrects the displacement (shift towards rehydration). Rats are intolerant of a swelling of intracellular fluids to a value above normal, in that if this does occur it inhibits the drinking tendency that is aroused by a simultaneous deficit in extracellular volume (Stricker 1973). Hence a signal that inhibits drinking arises from cell swelling, as shown in Box 2. By contrast, the system shows little or no sensitivity to moderate increases in extracellular fluid volume above normal (Corbit 1965); Box 1 shows this region to be associated with neither excitation nor inhibition.

Ignore for the moment the two other arrows at the summing junction and just consider that, if there is a sufficient excitatory signal and water is present, the animal drinks. Water drunk goes into the stomach and intestine, and drinking rate is integrated to give gut water volume. However, water entering the gut also tends to leave the gut and enter the blood. Can we characterise loss of water from the gut, in systems fashion? Figure 6.9b shows what happens after the stomach is artificially loaded with water (O'Kelly *et al.* 1958). The volume of water in the gut decays exponentially. Expressed mathematically, flow of water out of the gut is K_S (a constant) times volume. Hence there is in the control diagram a

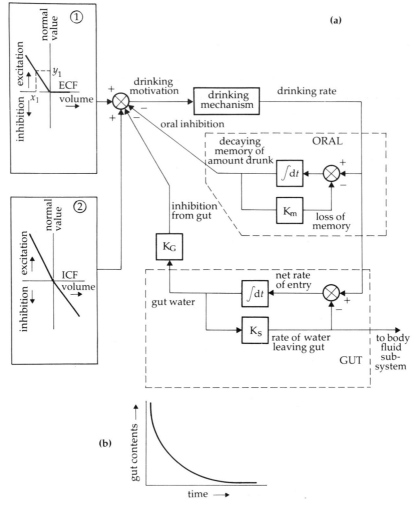

Fig. 6.9. (a) A model of drinking and (b) graph showing decay of gut water content following a load, as a function of time.

box containing K_S. Flow out of the gut is subtracted from flow in (drinking rate) to give net flow, and this term is then integrated to give gut water volume.

Models of drinking have been based on the assumption that both the amount of fluid in the gut and a memory of the amount drunk (arising from stimulation of oral receptors) inhibit further drinking (Toates & Oatley 1970). In fact, the system is much more complex than is suggested here (Toates 1980a) but Fig. 6.9a is a useful first approximation. We have a signal arising from gut water and exerting an inhibition on drinking (K_G is a constant affecting this). Turning to the section marked 'oral', we integrate drinking rate in order to give a memory (envisaged as neural activity) of the amount drunk. Drinking models include the assumption that this memory, and hence the associated inhibition, decays with time. This would seem logical: the animal's thirst ought not to be permanently inhibited because it has taken a drink. So a decay of memory is built in; K_m dictates the rate of this decay. This gives a decay of memory analogous to the decay of level of gut contents.

We have considered just two component subsystems of a complex model. Despite this complexity, one can build all of the system into a computer program and then question the computer to see what happens. For example, we can ask what would happen if the organism had no inhibition from the mouth and stomach on drinking. What use do these inhibitory pathways serve? Could drinking be switched off simply by a correction of the body fluid states that gave rise to it? To answer this, we would remove the inhibitory links from the program and might, for example, 'inject' the program with concentrated salt solution. In response to this treatment, the intact animal drinks just about enough to neutralise the salt load, as does the program when left complete. However, when inhibition from the mouth and gut is omitted, the program drinks 3–4 times as much as the real rat. This is due to time-lags in the system (e.g. delay in the gut). As well as such hypothetical questions, which would not be easy to test on a real animal, we can also determine how particular treatments, such as sodium depletion, affect the model. What we are asking is whether the features of the model can account for the way in which the animal behaves.

To a considerable extent, whether an animal drinks is determined by the state of its body fluids, though that is not to deny the importance of external stimuli. The nervous system detects body fluid states and, in conjunction with information on water in the environment, decides whether or not to drink. Many important aspects of drinking (e.g. how much an animal ingests following water deprivation) can be understood mainly in terms of its body fluid profile, and that is all that the model in Fig. 6.9a is able to achieve. The role of other factors is discussed by Toates (1981).

6.5   A model of sexual behaviour

Feeding and drinking are aspects of homeostatic mechanisms, the function of which is to maintain physiological states by supplying energy and water. Sexual behaviour serves a rather different end and no obvious physiological variable that is being regulated can be identified. However, that need not stop us from speculating as to the properties of the neural mechanisms that underlie sexual

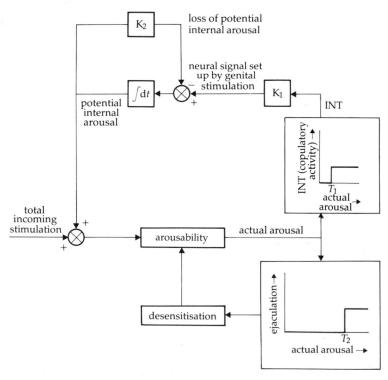

Fig. 6.10. Toates and O'Rourke's model of sexual behaviour in the male rat.

motivation, and this section shows how such speculation can be represented by models. We can ask questions similar to those mentioned above. Do animals really behave in the way that the model predicts? Is it necessary to include such-and-such a component to produce a realistic simulation of behaviour?

Toates and O'Rourke (1978) constructed a computer model of sexual behaviour in male rats. For reasons just indicated, the model is largely speculative and does not have the secure foundation in physiology that a drinking model can have. That is to say, in general, we do not have independent evidence for the existence of the components; they are there because the whole system appears to demand the inclusion of such components. The model shows the kind of processes we believe to exist, and it is constructed in the hope that it will suggest crucial experiments and give new theoretical insights. Figure 6.10 shows the model. We start with the assumption that sexual behaviour in the male is aroused by stimuli from the female or, in the appetitive phase, stimuli that in the past have been associated with females. Figure 6.10 shows an operator marked 'arousability'. This operator summarises the state of the male's nervous system as far as his proneness to sexual arousal is concerned. A sexually 'exhausted' male would have relatively low arousability, whereas a male that had been deprived of sexual contact would have relatively high arousability. Note that, put this way, arousability does not refer to an actual level of neural activity. Rather, it refers to the *capacity* of the nervous system to be aroused, given appropriate external stimulation. In Fig. 6.10, a variable termed 'total incoming stimulation' refers to stimuli (e.g. visual, olfactory, tactile) from the female that are able to arouse sexual behaviour (ignore for the moment 'potential internal arousal'). Whether or not such stimulation produces actual sexual arousal depends upon arousability in the following manner:

actual arousal = total incoming stimulation × arousability.

If actual arousal is sufficiently high then, assuming a receptive partner, the male initiates copulation. T_1 is a threshold; if actual arousal is above T_1 the rat performs a series of intromissions. If actual arousal is below T_1 then, as the vertical axis indicates, the animal does not attempt copulation. INT represents excitation from copulatory activity; this is either zero or a fixed level of stimulation from each of a series of discrete intromissions, which continues until arousal falls below T_1. It is assumed that each intromission sets up a neural signal at the genitals that travels to the brain. K_1 transforms the term INT into the neural signal that is set up by the associated genital stimulation. According to the model, bursts of neural activity arising from the intromissions sum to give a cumulative (integrated) level of activity within the nervous system. This is based upon the fact that ejaculation (described shortly) is only attained following several intromissions.

'Potential internal arousal' represents this integration of the bursts of activity. The model assumes that arousal is 'self-generating', i.e. arousal leads to intromission, and intromission in turn leads to further arousal. This loop is thus closed, and provides an example of positive feedback. Potential internal arousal and total incoming stimulation are summed. When actual arousal reaches a threshold value (T_2) that is considerably higher than T_1, the animal ejaculates. Activation of the ejaculatory mechanism causes arousability to be lowered. The time that the organism takes to recover from the loss of arousability will vary from species to species. In rats, a series of ejaculations occurs before long-term satiety is shown (Sachs & Barfield 1976). Each ejaculation appears to lower sensitivity, but then a fast recovery occurs. After a few ejaculations, however, the animal is unable to recover until a relatively long period of time has elapsed.

Suppose an animal takes eight intromissions to achieve one ejaculation. On another occasion, under the same conditions of prior deprivation, it is allowed three intromissions, a pause in mating of 1 h is then imposed by the experimenter, and the rat allowed to complete its copulation to ejaculation. Following the pause, eight rather than five intromissions may be needed to attain ejaculation. Our interpretation is that stored excitation dissipates as a function of time. It appears to do so at a rate that is proportional to the amount of excitation remaining. To accommodate this the model includes the constant term K_2, which reduces potential internal arousal according to the rate of loss of excitation. The loss is subtracted from the excitatory signal.

What is the value of a model such as this? It does not show how known neurophysiological components interact to generate behaviour, since much of it is derived from the behaviour itself. However, it does represent a theory that is compatible with our understanding so far. It provides an organising framework, and its terms are fairly unambiguous, so that the model can easily be modified (or discarded!) in the light of new developments. The model invites discussion of what its components represent. For instance, do arousability and its loss depend at all upon accumulation and depletion of seminal fluid? The evidence does not favour this (Toates 1980b); probably a better guess is that somewhere within the CNS a neural circuit, or an individual synapse, is desensitised at ejaculation (Lorenz's process of losing fluid from a tank is analogous to this). In terms of the model, this consists of a change in a single parameter, arousability. The model includes the recovery of arousability as a feature leading to changes that occur as the period of prior deprivation of sexual contact increases. In rats the following such changes are observed: (1) the animal becomes easy to arouse, i.e. the latency to the initial mounting is short; (2) ejaculation becomes easy to obtain (relatively few intromissions); (3) many ejaculations occur before sexual satiety sets in; (4) discrimination becomes relatively low (a wide range of stimuli may be adequate to arouse mating).

Further questions need to be answered. For example, to what extent can the model be generalised to other species? How realistic is it, even for the rat, in anything other than a simple laboratory situation? Nevertheless, construction of a theoretical model of this kind is justified as a 'first attempt' on the grounds that there are so many data available for laboratory rats.

6.6   Goal-directed behaviour

Model building is closely associated with the theory of *servomechanisms*, which considers how complex goal-directed systems are controlled. Related to the goal concept, behaviour is often described as having *purpose*. By this it is meant that behaviour is directed towards attaining certain goals. As I shall attempt to demonstrate, formal models, borrowed from engineering, strip some of the mystery from purposive behaviour. They provide a mechanistic realisation of how such behaviour arises. There are at least two aspects to this problem, one of which has been described already. Take drinking as an example. Depletion of body fluids threatens survival, so the organism is equipped with mechanisms whereby dehydration of body fluids raises drinking motivation. The feedback aspect is clear and the goal or purpose of this mechanism can be seen to be maintenance of optimal body fluid states. So far we have considered water to be freely available in the rat's cage, but for an understanding of goal-directedness we must consider the behaviour of the dehydrated animal before it reaches water. In the wild, it may move towards a known source of water. This behaviour is goal-directed, the goal being a particular place. The animal might use any of a variety of routes to reach the goal. The following example can be used to bring out more clearly the control theory or servomechanism aspect of this process.

Start by considering the following simple problem (Powers 1978). A bird is sitting on its nest and an experimenter surreptitiously tilts the nest through an angle θ (we must assume that the bird does not fly away!). The bird would typically tilt by an angle of only $\frac{1}{2}\theta$. In other words, the bird takes corrective action that compensates to some extent for the disturbance. This implies that the bird perceives the disturbance, because otherwise it could not take corrective action. That is to say, it has some inner representation of where it *ought* to be. Engineers call this a *set-point,* a term that means much the same as goal. Somewhere in the nervous system, the position that the bird actually occupies is compared with the set-point. If an error is detected, then action, in this case muscular effort and weight redistribution, is taken in order to minimise the disturbance. Figure 6.11 shows a model of the bare essentials of this process.

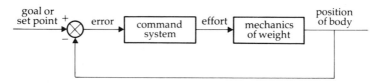

Fig. 6.11. A model of the system used by a bird in maintaining body position in its nest. The goal or set-point of the system is set by the nervous system. The actual position of the body is compared with this. Any difference is perceived as an error, and the command system produces a new muscular effort. This changes the position of the body. Note that action can be instigated in either of two ways. An external disturbance can create an error or the bird can change the goal, for example it may decide to peer over the edge of the nest.

Having introduced the idea of the set-point, let us relate it to earlier motivation models by reconsidering the water-deprived animal moving towards a *particular* location of water. Suppose that the site is distant and, at the time in question, the water cannot be perceived by the animal. What does moving there entail, in terms of mechanisms within the animal's nervous system? First, it involves an inner representation of the environment, a spatial or cognitive map. The idea that animals are able to construct maps of their environment has gained support from neurophysiological work (O'Keefe & Nadel 1978); but having a map within the nervous system does not, on its own, instruct the organism on how to react to that environment. The animal is in danger of being left buried in thought! In practice, different locations within the map must form the goal of behaviour at different times. For example, at times of dehydration, the goal must be a source of water. Body fluid detectors need to make contact with points on the map. A model of how this might be achieved was proposed by Deutsch (1960). Although the details of such models are beyond our scope here, the principle of establishing a cognitive map, a point on which then forms the goal of behaviour, is highly relevant. The animal has to move in such a way as to minimise the distance between itself and the goal; achieving this is another example of a negative feedback system.

6.7   Doing only one thing at a time—how to classify motivational systems

Performance of one activity is, of course, often incompatible with simultaneous performance of another. An obvious example is that an animal fleeing from a predator must abandon feeding. In general, at any point in time, an animal may have several *candidates* for behavioural expression. For example, it may simultaneously be low in energy, dehydrated and in the presence of a potential mate. As described in Chapter 5, the animal must be equipped with mechanisms for decision making, in order to ensure that it does not get stuck trying to do two incompatible things at once. Construction of models is a useful way of exploring the kind of processes that are involved.

We can quantify this in the following way. At a particular moment, an animal has '10 g' of hunger and '10 ml' of thirst, meaning that, if given the opportunity, it would consume food and water in these quantities. If only *one* of the two commodities was present then the motivation appropriate to this commodity would take command and cause ingestion. But when both commodities are present, if the animal starts by eating then the tendency to drink must be inhibited for the duration of the meal. Exactly how this might be achieved has been a subject of speculation by ethologists interested in a modelling approach. Some interesting ideas have been put forward by Ludlow (1980) and the essence of his approach is summarised in Fig. 6.12 and its caption.

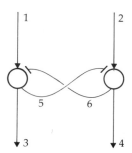

Fig. 6.12. Ludlow's model of competition. Imagine two neurons 3 and 4 that cannot fire simultaneously. Neuron 3 is excited by 1, and 4 is excited by 2. When 1 is able to activate 3 then inhibition is exerted via connection 5. This means that even if 2 is exciting 4, 4 is unable to fire. After a while the strength of 1 may fall, in which case 5 exerts less inhibition; 4 may therefore be able to fire. This exerts inhibition via 6, which means 3 is unable to fire. In other words, despite signals at 1 *and* 2, only 3 *or* 4 can fire. Ludlow proposes that the neurons might represent hunger (say, neuron 1), thirst (2), feeding (3) and drinking (4). To what extent the *actual* neural embodiment of motivational competition is as shown is open to discussion.

Figure 6.13 shows a development of this model. The animal has tendencies to explore, groom, mate, drink, eat, etc., each of which is, at any instant, a candidate to take command of behaviour. The motivation with the strongest tendency takes command. Suppose that eating and drinking are the two strongest candidates for expression. The strengths of the drinking and eating commands are each 10 units. Assume that these two values give almost equal tendencies, but that by a marginal difference the drinking candidate wins. The motor program for drinking is switched on. The strength of the command is 10 units. This exerts inhibition on the eating tendency, as shown by the interaction. The drinking command is multiplied by a factor of 1.5 to give the strength of inhibition on feeding. Thus, an influence of −15 is exerted on feeding, which totally cancels out the 10 unit feeding command. In effect, the feeding tendency is −5, so an otherwise strong candidate becomes a weak one. As thirst is reduced by water ingestion there will come a point at which the inhibition it exerts becomes inadequate to restrain feeding. Hunger will then oust thirst as the dominant tendency and the animal will eat. How long the animal persists with an activity, once initiated, will depend upon the strength of inhibition that it exerts on other potential activities. In this example, if the gain of 1.5 were instead to be 15, drinking would be much more likely to persist.

A discussion of competition, and particularly the construction of formal models of the process, raises some fundamental issues in motivation theory. For example, at what level in the processing of information and in the implementation of responding does competition occur? Ludlow's model suggests that the current activity cancels the effect of other motivations. This may be the level at which interaction occurs, but we do not yet know. Much of this is speculation, as we lack understanding of the way in which the real

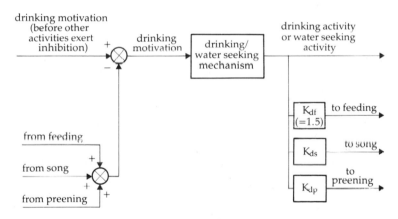

Fig. 6.13. Model of motivational interaction as proposed by Ludlow (1980). Consider drinking motivation to arise from factors intrinsic to this controller (e.g. body water, cue strength of water, etc.). Assume that the animal is drinking; inhibition is exerted on all potentially competing activities. For instance, K_{df} (assumed to be 1.5) is a term relating drinking activity to the inhibition drinking exerts on feeding. Other inhibitory links are made to song, preening, etc. The mechanism whereby other activities (e.g. feeding) can exert inhibition on the effect of drinking motivation is also shown. If such inhibition is exerted then, in effect, drinking motivation is lowered, at least insofar as its candidature for behavioural expression is concerned. The summing junction at the bottom left indicates a final common path for the various inhibitory influences on drinking. Note that inhibition comes only from whichever of these is being performed.

nervous system makes such decisions. Toates (1981) raised the possibility that, rather than motivational states, it might be *goals* that compete. Consider the following example. Suppose at a choice point, an animal turns left to arrive at a sleeping site and right to obtain both food *and* water. It comes to the choice point short of both food and water, turns right, and starts to feed. One might argue that hunger was stronger than the tendency to go to the sleeping site and was also stronger than thirst. But is one necessarily to assume that only hunger contributed to the strength of the tendency to turn right? Rather than competing, the component tendencies arising from hunger and thirst, each prompting the same response, might *add* their effects at the choice point. In these terms, if food and water were at *different* sites then the two potential goals would compete, as would be the case in this example when the animal gets to the goal-box. We do not yet know how animals react in such circumstances, but the question is surely one that is worth raising.

These considerations in turn raise another question. How are we to classify motivations? Suppose that, when an animal is either feeding or drinking, there is inhibition of the other activity at the level of motivation. It would seem logical to conclude that the animal, in the process of ingesting a substance, is under the control of *either* hunger *or* thirst. This may be so in the laboratory, looking at rats eating dry food and with free water available, but what about a hungry and thirsty gerbil eating a moist plant in the desert? Does it do so under the control of a hunger or of a thirst system? Can we make any such distinctions? Surely, both energy and fluid depletion would motivate the animal. Hence, it is likely that for a given activity more than one motivational system can be involved. In this context, an interesting case is described by Bernays (1977) for the locust. The locust normally obtains water from its food in excess of requirements. If, however, only very dry food has been available, the locust will accept water or will widen the choice of plants that it will eat. This suggests that normal ingestion of moist plants is under the control of both hunger *and* thirst systems.

6.8   Concluding remarks

Various models have been described in this chapter. Whatever the type of model, in practice the theorist finds himself oscillating between experiments and data on the one hand and modelling on the other. The model, often built into a computer program, is used to sort out ideas, generate hypotheses and test the outcomes of different assumptions. These are then tested by collecting experimental data. Sometimes the model reveals unexpected complexities and at other times suggests that a few simple assumptions *may* be adequate (as with Bolles' fly feeding model). The models described here have ranged from those in which speculation was minimal, components being justified from physiological data (e.g. the body-fluid exchange system) to the much more speculative (e.g. the sex model or Ludlow's competition model).

The models described at the outset considered the properties of *systems*, and how models can capture the essence of a system by showing the mode of interconnection of its components. Models are invariably simplifications; a model involves only aspects of a real system. For example, we may simplify the system by taking for granted the presence of food, water or a mate, and hence avoid problems associated with how the animal locates such goal objects (e.g. the sex model of Toates and O'Rourke). A system based upon such a simplification can show how internal states contribute to motivation. By contrast, another model might take detection of internal states for granted and attempt to show how, given a particular deficit, the animal implements appropriate goal-directed behaviour. Another simplification is to model distinct systems (e.g. feeding or drinking), ignoring the powerful interactions between these systems (Toates 1979b). Yet one of the major determinants of water intake is food intake. This raises the problem of how we classify motivations, which is especially acute when discussing competition.

A recurrent theme in this chapter has been the importance of both internal and external factors in motivation. Traditionally, a dichotomy has been made between behaviour patterns like feeding, in which internal factors were considered to be all-important, and those such as sex, in which external factors were thought paramount. Models can help to take us beyond the oversimplification inherent in such a dichotomy. Whereas we recognise that in feeding and drinking internal states play a vital physiological role not apparent in the case of sex, models have helped to demonstrate the great importance of external factors in feeding (e.g. the models outlined by Bolles and by McFarland and Sibly), and of internal factors in sexual behaviour (e.g. the model proposed by Toates and O'Rourke).

No discussion of modelling can be complete without some reference to the controversy that surrounds use of the term 'set-point'. Figure 6.11 showed a set-point as an integral part of a model. The actual state is compared with the set-point and, when there is an error, action is instigated. The set-point is seen to have physical identity, for example a particular pattern or level of neural activity. However, one can compare this situation with that shown in Fig. 6.7b. Here, when blood glucose level falls below value c_1 corrective action, feeding, occurs. There is no actual signal corresponding to the level c_1, against which concentration is compared. Nonetheless, because to all intents and purposes it is *as if* c_1 forms a set-point, some would indeed use that term. But this can be deceptive. The relative stability of body weight in Bolles' fly model might lead one to assume that weight is compared with a set-point (in the soles of the fly's feet perhaps!), whereas this body weight is actually the outcome of such factors as food availability and inhibition from the gut. As several models may yield the same result, caution is clearly in order. A point made at the start of the chapter may usefully be reiterated here: while models are helpful, one should not become too fond of any particular one.

6.9   Selected reading

A basic introduction to the mathematics of control theory and model building is given by Toates (1975), and McFarland (1971) provides an account of the application of control theory to behaviour which demands some mathematical understanding. An introduction to motivation written in terms of model building, and discussing hunger, thirst, sex, temperature control, etc., is to be found in the book by Toates (1980). An account of the work of McFarland and associates in this area is given by McFarland and Houston (1981), in a book which again assumes some familiarity with mathematics. For some of the controversy and debate in this area the reader might consult the chapters by Bolles, Davis and Hogan in the volume edited by Toates and Halliday (1980).

REFERENCES

The section(s) in which each reference appears is given after the reference.

Alexander R.D. (1960) Sound communication in Orthoptera and Cicadidae. In: *Animal Sounds and Communication* (ed. W.E. Lanyon & W.N. Tavolga), pp. 38–91. American Institute of Biological Sciences, Washington.
1.2.3

Alexander R.McN. (1982) *Optima for Animals*. Edward Arnold, London.
5.8, 5.9

Amsel A. & Roussel J. (1952) Motivational properties of frustration: I. Effect on a running response of the addition of frustration to the motivational complex. *Journal of Experimental Psychology* 42, 363–368.
5.4

Andrew R.J. (1956) Normal and irrelevant toilet behaviour in *Emberiza* spp. *Animal Behaviour* 4, 85–91.
5.6.1

Andrew R.J. (1974) Arousal and the causation of behaviour. *Behaviour* 51, 135–165.
4.5

Andrew R.J. (1976) Attentional processes and animal behaviour. In: *Growing Points in Ethology* (ed. P.P.G. Bateson & R.A. Hinde), pp. 95–133. Cambridge University Press, Cambridge.
4.5.1

Angevine J.R. Jr & Cotman C.W. (1981) *Principles of Neuroanatomy*. Oxford University Press, Oxford.
2.8

Archer J. (1970) Effects of population density on behaviour in rodents. In: *Social Behaviour in Birds and Mammals* (ed. J.H. Crook), pp. 169–210. Academic Press, London.
4.6

Archer J. (1976) The organisation of aggression and fear in vertebrates. In: *Perspectives in Ethology*, Vol. 2 (ed. P.P.G. Bateson & P.H. Klopfer), pp. 231–298. Plenum Press, New York.
4.6

Arnold S.J. (1976) Sexual behaviour, sexual interference and sexual defence in the salamanders *Ambystoma maculatum*, *Ambystoma tigrinum* and *Plethodon jordani*. *Zeitschrift für Tierpsychologie* 42, 247–300.
4.3.2

Aronson L.R. (1971) Further studies on orientation and jumping behavior in the gobiid fish, *Bathygobius soporator*. *Annals of the New York Academy of Sciences* 188, 378–392.
2.6

Atkinson J.W. & Birch D. (1970) *The Dynamics of Action*. John Wiley & Sons, New York.
5.6, 5.6.2

Autrum H., Jung R., Loewenstein W.R., MacKay D.M. & Teuber H.-L. (eds) (1971–1981) *Handbook of Sensory Physiology* Vols I to XII. Springer-Verlag, Berlin.
1.4

Baerends G.P., Brouwer R. & Waterbolk H.T. (1955) Ethological studies on *Lebistes reticulatus* (Peters): I. An analysis of the male courtship pattern. *Behaviour* **8**, 249–334.
4.3.1

Bangert H. (1960) Untersuchungen zur Koordination der Kopf- und Bein-bewegungen beim Haushuhn. *Zeitschrift für Tierpsychologie* **17**, 143–164.
3.2.2

Barfield R.J. & Sachs B.D. (1968) Sexual behavior: stimulation by painful electric shock to skin in male rats. *Science* **161**, 392–395.
4.5

Barlow G.W. (1968) Ethological units of behavior. In: *The Central Nervous System and Fish Behavior* (ed. D. Ingle), pp. 217–232. University of Chicago Press, Chicago.
3.2.1

Barlow G.W. (1977) Modal action patterns. In: *How Animals Communicate* (ed. T.A. Sebeok), pp. 98–134. Indiana University Press, Bloomington.
3.2.1, 3.5

Barlow H.B. (1972) Single units and sensation: a neuron doctrine for perceptual psychology? *Perception* **1**, 371–394.
Introduction

Barlow H.B. (1981) Ferrier Lecture: critical limiting factors in the design of the eye and visual cortex. *Proceedings of the Royal Society of London* B **212**, 1–34.
2.8

Barnett S.A. & Cowan P.E. (1976) Activity, exploration, curiosity and fear: an ethological study. *Interdisciplinary Science Reviews* **1**, 43–62.
4.6

Bastock M. & Manning A. (1955) The courtship of *Drosophila melanogaster*. *Behaviour* **8**, 85–111.
5.3.1

Bastock M., Morris D. & Moynihan M. (1953) Some comments on conflict and thwarting in animals. *Behaviour* **6**, 66–84.
5.6.1

Bateson P.P.G. (1983) Genes, environment and the development of behaviour. In: *Animal Behaviour, Vol. 3: Genes, Development and Learning* (ed. T.R. Halliday & P.J.B. Slater). Blackwell Scientific Publications, Oxford.
3.2.1

Beach F.A. & Whalen R.E. (1959) Effects of intromission without ejaculation on sexual behaviour in male rats. *Journal of Comparative and Physiological Psychology* **52**, 476–481.
4.6.1, 5.6.2

Bentley D. (1977) Control of cricket song patterns by descending interneurons. *Journal of Comparative Physiology* **116**, 19–38.
3.2.1

Bentley D. & Konishi M. (1978) Neural control of behavior. *Annual Review of Neuroscience* **1**, 35–60.
3.2.1, 3.5

Berg H.C. (1975) How bacteria swim. *Scientific American* **233** (2), 36–44.
2.2

Berg H.C. & Purcell E.M. (1977) Physics of chemo-reception. *Biophysics Journal* **20**, 193–219.
2.2

Bernays E.A. (1977) The physiological control of drinking behaviour in nymphs of *Locusta migratoria*. *Physiological Entomology* **2**, 261–273.
6.7

Bernstein N. (1967) *The Coordination and Regulation of Movements*. Pergamon Press, Oxford.
2.9

Berthold P. (1974) Circannual rhythms in birds with different migratory habits. In: *Circannual Clocks* (ed. E.T. Pengelley), pp. 55–94. Academic Press, New York.
3.3.3

Bizzi E., Kalil R.E. & Tagliasco V. (1971) Eye–head coordination in monkeys: evidence for centrally patterned organisation. *Science* **173**, 452–454.
2.9

Bizzi E., Polit A. & Morasso P. (1976) Mechanisms underlying achievement of final head position. *Journal of Neurophysiology* **39**, 435–444.
2.10.2

Blakemore R.P., Frankel R.B. & Kalmijn Ad. J. (1980) South-seeking magnetotactic bacteria in Southern Hemisphere. *Nature* (London) **286**, 384–385.
2.2

Blakemore R.P. & Frankel R.B. (1981) Magnetic navigation in bacteria. *Scientific American* **245** (6), 42–49.
2.3

Bolles R.C. (1975) *Theory of Motivation*, 2nd edn. Harper & Row, New York.
4.5, 4.6.1, 4.8, 5.4, 5.6.2, 5.8.2

Bolles R.C. (1980) Some functionalistic thoughts about regulation. In: *Analysis of Motivational Processes* (ed. F.M. Toates & T.R. Halliday), pp. 63–75. Academic Press, London.
4.6.1, 4.8, 6.3.1, 6.8

Bonsall R.W., Zumpe D. & Michael R.P. (1978) Menstrual cycle influences on operant behaviour of female rhesus monkeys. *Journal of Comparative and Physiological Psychology* **92**, 846–855.
4.3.3

Booth D.A. (1972) Conditioned satiety in the rat. *Journal of Comparative and Physiological Psychology* **81**, 457–571.
4.6.1

Booth D.A. (1978) *Hunger Models: Computable Theory of Feeding Control*. Academic Press, London.
6.3.4

Booth D.A. (1980) Conditioned reactions in motivation. In: *Analysis of Motivational Processes* (ed. F.M. Toates & T.R. Halliday), pp. 77–102. Academic Press, London.
6.3.4

Booth D.A. & Toates F.M. (1974) A physiological control theory of food intake in the rat: Mark 1. *Bulletin of the Psychonomic Society* **3**, 442–444
6.3.4

Brady J. (1975) Circadian changes in central excitability—the origin of behavioural rhythms in tsetse flies and other animals? *Journal of Entomology* (A) **50**, 79–95.
4.5

Brady J. (1979) *Biological Clocks*. Edward Arnold, London.
4.5

Brady J. (ed.) (1982) *Biological Timekeeping*. Cambridge University Press, Cambridge.
4.5.2

Brandt T., Dichgans J. & Koenig E. (1973) Differential effects of central versus peripheral vision on egocentric and exocentric motion perceptions. *Experimental Brain Research* **14**, 476–491.
2.5

Brockmann H.J. (1980) The control of nest depth in a digger wasp (*Sphex ichneumoneus* L.) *Animal Behaviour* **28**, 426–445.
3.4.1, 4.4

Brown R. & McFarland D.J. (1979) Interaction of hunger and sexual motivation in the male rat: a time sharing approach. *Animal Behaviour* **27**, 887–896.
5.6.2

Bulloch A.G.M. & Dorsett D.A. (1979) The functional morphology and motor innervation of the buccal mass of *Tritonia hombergi*. *Journal of Experimental Biology* **79**, 7–22.
3.2.1

Bullock T.H. (1976) In search of principles of neural integration. In: *Simpler Networks and Behavior* (ed. J.C. Fentress), pp. 52–60. Sinauer, Sunderland, Mass.
3.1

Busnel R.-G. & Fish J.F. (eds) (1980) *Animal Sonar Systems*. Plenum Press, New York.
1.2.3

Camhi J.M. (1974) Neural mechanisms of response modification in insects. In: *Experimental Analysis of Insect Behaviour* (ed. L. Barton-Browne), pp. 60–86. Springer-Verlag, Berlin.
1.1

Cane V. (1978) On fitting lower order Markov chains to behaviour sequences. *Animal Behaviour* **26**, 332–338.
3.3.2

Capranica R.R. (1976) Morphology and physiology of the auditory system. In: *Frog Neurobiology* (ed. R. Llinás & W. Precht), pp. 551–575. Springer-Verlag, Berlin.
1.2.3

Carlson J.R. & Bentley D. (1977) Ecdysis: neural orchestration of a complex behavioral performance. *Science* **195**, 1006–1008.
3.3.1

Cartwright B.A. & Collett T.S. (1982) How honey bees use landmarks to guide their return to a food source. *Nature* (London) **295**, 560–564.
2.5

Cheng M.-F. (1979) Progress and prospects in ring dove research: a personal view. *Advances in the Study of Behavior* **9**, 97–129.
4.5.1

Cizek L.J. (1961) Relationship between food and water ingestion in the rabbit. *American Journal of Physiology* **201**, 557–566.
5.5.1

Cohen S. & McFarland D.J. (1979) Time-sharing as a mechanism for control of behaviour sequences during the courtship of the three-spined stickleback *Gasterosteus aculeatus*. *Animal Behaviour* **27**, 270–283.
5.6.2

Collett T.S. (1982) Do toads plan routes? A study of the detour behaviour of *Bufo viridis*. *Journal of Comparative Physiology* **146**, 261–271.
2.6

Collett T.S. & Land M.F. (1975) Visual control of flight behaviour in the hoverfly, *Syritta pipiens* L. *Journal of Comparative Physiology* **99**, 1–66.
2.5

Collett T.S. & Land M.F. (1978) How hoverflies compute interception courses. *Journal of Comparative Physiology* **125**, 191–204.
2.5

Collias N.E. & Collias E.C. (1962) An experimental study of the mechanism of nest-building in a weaverbird. *Auk* **79**, 568–595.
3.4.1

Corbit J.D. (1965) Effect of intravenous sodium chloride on drinking in the rat. *Journal of Comparative and Physiological Psychology* **60**, 397–406.
6.4

Cowie R.J., Krebs J.R. & Sherry D.F. (1981) Food storing by marsh tits. *Animal Behaviour* **29**, 1252–1259.
3.4.1

Crook J.H. (1964) Field experiments on the nest construction and repair behaviour of certain weaverbirds. *Proceedings of the Zoological Society of London* **142**, 217–255.
3.4.1

Cynader M. & Berman N. (1972) Receptive-field organisation of monkey superior colliculus. *Journal of Neurophysiology* **35**, 187–201.
2.10.1

Daan S. (1981) Adaptive daily strategies in behavior. In: *Handbook of Behavioral Neurobiology*, Vol. 4, *Biological Rhythms* (ed. J. Aschoff), pp. 275–298. Plenum Press, New York.
4.5.2

Daly H.B. (1969) Learning of a hurdle jump response to escape cues paired with reduced reward or frustrative nonreward. *Journal of Experimental Psychology* **76**, 146–157.
5.4

Dane B., Walcott C. & Drury W.H. (1959) The form and duration of the display actions of the goldeneye (*Bucephala clangula*). *Behaviour* **14**, 265–281.
3.2.2

Davis W.J. (1976) Organisational concepts in the control motor organisation of invertebrates. In: *Neural Control of Locomotion* (ed. R. Herman, S. Grillner, P.S.G. Stein & D. Stuart), pp. 265–292. Plenum Press, New York.
5.2

Davis W.J., Mpitsos G.J., Pinneo J.M. & Ran J.L. (1977) Modification of the behavioural hierarchy of *Pleurobranchaea*. I: Satiation and feeding mechanisms. *Journal of Comparative Physiology* **117**, 99–125.
5.2

Dawkins M. (1974) Behavioural analysis of coordinated feeding movements in the gastropod *Lymnaea stagnalis*. *Journal of Comparative Physiology* **92**, 255–271.
3.3.3

Dawkins M. & Dawkins R. (1974) Some descriptive and explanatory stochastic models of decision-making. In: *Motivational Control Systems Analysis* (ed. D.J. McFarland). Academic Press, London.
3.2.2, 5.3.1, 5.3.2

Dawkins R. (1969) The attention threshold model. *Animal Behaviour* **45**, 83–103.
5.3.1

Dawkins R. (1976) Hierarchical organisation: a candidate principle for ethology. In: *Growing Points in Ethology* (ed. P.P.G. Bateson & R.A. Hinde), pp. 7–54. Cambridge University Press, Cambridge.
3.4.2, 5.1.1, 5.2

Dawkins R. & Dawkins M. (1973) Decisions and the uncertainty of behaviour. *Behaviour* **45**, 83–103.
3.2.2, 5.6.2

Delcomyn F. (1980) Neural basis of rhythmic behavior in animals. *Science* **210**, 492–498.
3.2.1

Denton E. (1971) Reflectors in fishes. *Scientific American* **224** (1), 64–72.
1.2.1

Dethier V.G. (1966) Insects and the concept of motivation. *Nebraska Symposium on Motivation* 105–136.
5.2

Dethier V.G. (1974) The specificity of the labellar chemoreceptors of the blow fly and the response to natural foods. *Journal of Insect Physiology* **20**, 1859–1869.
1.2.5

Dethier V.G. (1976) *The Hungry Fly: A Physiological Study of the Behavior Associated with Feeding*. Harvard University Press, Cambridge, Mass.
6.3.1

Deutsch J.A. (1960) *The Structural Basis of Behavior*. University of Chicago Press, Chicago.
6.6

Dorsett D.A., Willows A.O.D. & Hoyle G. (1973) The neuronal basis of behavior in *Tritonia*. IV. The central origin of a fixed action pattern demonstrated in the isolated brain. *Journal of Neurobiology* **4**, 287–300.
3.2.1

Dräger U.C. & Hubel D.H. (1975) Responses to visual stimulation and relationship between visual, auditory and somatosensory inputs in mouse superior colliculus. *Journal of Neurophysiology* **38**, 690–713.
2.8

Eckert R., Naitoh Y. & Machemer H. (1976) Calcium in the bioelectric and motor functions of *Paramecium*. *Symposium of the Society for Experimental Biology* **30**, 233–255.
2.4

Eckert R. & Randall D. (1978) *Animal Physiology*. W.H. Freeman, San Francisco.
2.4

Eisner T., Silberglied R.E., Aneshansley D., Carrel J.E. & Howland H.C. (1969) Ultraviolet video-viewing: the television camera as an insect eye. *Science* **166**, 1172–1174.
1.2.1

Elsner N. (1973) The central nervous control of courtship behaviour in the grasshopper *Gomphocerippus rufus* (L.). In: *Neurobiology of Invertebrates*, pp. 261–287. Hungarian Academy of Sciences, Tihany.
3.2.1

Emlen S. (1975) The stellar orientation system of a migratory bird. *Scientific American* **233** (2), 102–111.
1.1

Epstein A.N., Fitzsimons J.T. & Rolls B.J. (1970) Drinking induced by injection of angiotensin into the brain of the rat. *Journal of Physiology* **210**, 457–474.
5.5.2

Erickson C.J. & Martinez-Vargas M.C. (1975) The hormonal basis of cooperative nest-building. In: *Neural and Endocrine Aspects of Behaviour in Birds* (ed. P. Wright, P.G. Caryl & D.M. Vowles), pp. 91–109. Elsevier, Amsterdam.
4.5.1

Ewert J.-P. (1974) The neural basis of visually guided behavior. *Scientific American* **230** (3), 34–42.
Introduction

Ewert J.-P. (1980) *Neuroethology*. Springer-Verlag, Berlin.
2.5, 2.6

Fagen R.M. & Young D.Y. (1978) Temporal patterns of behavior: durations, intervals, latencies, and sequences. In: *Quantitative Ethology* (ed. P.W. Colgan), pp. 79–114. John Wiley & Sons, New York.
3.3.2

Falk J.L. (1971) The nature and determinants of adjunctive behavior. *Physiology and Behavior* **6**, 577–588.
5.4

Feekes F. (1972) 'Irrelevant' pecking in agonistic situations in Burmese red jungle fowl (*Gallus gallus spadiceus*). *Behaviour* **43**, 186–326.
5.6.1

Fentress J.C. (1972) Development and patterning of movement sequences in inbred mice. In: *The Biology of Behavior* (ed. J. Kiger), pp. 83–132. Oregon State University Press, Cornwallis.
5.4

Fentress J.C. (1976) Dynamic boundaries of patterned behaviour: interaction and self-organization. In: *Growing Points in Ethology* (ed. P.P.G. Bateson & R.A. Hinde), pp. 135–169. Cambridge University Press, Cambridge.
3.2.1, 5.4

Fitzsimons J.T. & Oatley K. (1968) Additivity of stimuli for drinking in rats. *Journal of Comparative and Physiological Psychology* **66**, 450–455.
6.4

Fitzsimons J.T. & Le Magnen J. (1969) Eating as a regulatory control of drinking. *Journal of Comparative and Physiological Psychology* **67**, 273–283.
5.5.2

Fitts P.M. & Posner M.I. (1973) *Human Performance*. Prentice Hall, London.
2.9

Fraenkel G.S. & Gunn D.L. (1961) *The Orientation of Animals*. Dover, New York.
1.2.1

Frey D.F. & Pimental R.A. (1978) Principal component analysis and factor analysis. In: *Quantitative Ethology* (ed. P.W. Colgan), pp. 219–246. John Wiley & Sons, New York.
5.3.2

Frisch K. von (1962) Dialects in the language of bees. *Scientific American* **207** (2), 79–87.
1.2.5

Gamow R.I. & Harris J.F. (1973) The infrared receptors of snakes. *Scientific American* **228** (5), 94–100.
1.2.1

Gelperin A., Chang J.J. & Reingold S.C. (1978) Feeding motor program in *Limax*. I. Neuromuscular correlates and control by chemosensory input. *Journal of Neurobiology* **9**, 285–300.
3.2.1

Georgopoulos A.P., Kalaska J.F. & Massey J.T. (1981) Spatial trajectories and reaction times of aimed movements: effects of practice, uncertainty and change in target location. *Journal of Neurophysiology* **46**, 725–743.
2.9

Gerhardt H.C. (1983) Communication and the environment. In: *Animal Behaviour, Vol. 2: Communication* (ed. T.R. Halliday & P.J.B. Slater), pp. 82–113. Blackwell Scientific Publications, Oxford.
1.2.3

Getting P.A. (1977) Neuronal organisation of escape swimming in *Tritonia*. *Journal of Comparative Physiology* **121**, 325–342.
3.4.2

Gibson J.J. (1950) *The Perception of the Visual World*. Houghton Mifflin, Boston.
1.2.1

Gleeson R.A. (1978) Functional adaptations in chemosensory systems. In: *Sensory Ecology* (ed. M.A. Ali), pp. 291–317. Plenum Press, New York.
1.2.5

Gogala M. (1978) Ecosensory functions in insects. In: *Sensory Ecology* (ed. M.A. Ali), pp. 123–153. Plenum Press, New York.
1.2.3

Gonshor A. & Melvill Jones G. (1973) Changes of human vestibulo-ocular response induced by vision reversal during head rotation. *Journal of Physiology* **234**, 102–103.
2.11

Goodman L.J. (1965) The role of certain optomotor reactions in regulating the stability in the rolling plane during flight in the desert locust *Schistocerca gregaria*. *Journal of Experimental Biology* **42**, 382–407.
2.5

Gould S.J. & Lewontin R.C. (1979) The spandrels of San Marco and the Panglossian paradigm: a critique of the adaptationist program. *Proceedings of the Royal Society of London* B **205**, 581–598.
5.8.2

Greene P.H. (1972) Problems of organization of motor systems. In: *Progress in Theoretical Biology, Vol. 2* (ed. R. Rosen & F.M. Snell), pp. 303–338. Academic Press, New York.
2.9

Griffin D.R. (1958) *Listening in the Dark: The Acoustic Orientation of Bats and Men*. Yale University Press, New Haven.
2.6

Grillner S. (1975) Locomotion in vertebrates: central mechanisms and reflex interaction. *Physiological Reviews* **55**, 247–304.
3.2.1

Grillner S. & Wallen P. (1977) Is there a peripheral control of the central pattern generators for swimming in dogfish? *Brain Research* **127**, 291–295.
3.2.1

Gwinner E. (1972) Endogenous timing factors in bird migration. In: *Animal Orientation and Navigation* (ed. S. Galler, K. Schmidt-Koenig, G.J. Jacobs & R.E. Belleville), pp. 321–338. NASA, Washington DC.
3.3.3

Gwinner E. (1981) Circannual systems. In: *Handbook of Behavioural Neurobiology*, Vol. 4, *Biological Rhythms* (ed. J. Aschoff), pp. 391–410. Plenum Press, New York.
4.5.2

Hailman J.P. & Sustare B.D. (1973) What a stuffed toy tells a stuffed shirt. *BioScience* **23**, 644–651.
3.3.2

Halliday T.R. (1974) Sexual behaviour of the smooth newt, *Triturus vulgaris* (Urodela: Salamandridae). *Journal of Herpetology* **8**, 277–292.
4.3.2

Halliday T.R. (1975) An observational and experimental study of sexual behaviour in the smooth newt, *Triturus vulgaris* (Amphibia: Salamandridae). *Animal Behaviour* **23**, 291–322.
3.4.1, 4.4

Halliday T.R. (1976) The libidinous newt. An analysis of variations in the sexual behaviour of the male smooth newt, *Triturus vulgaris*. *Animal Behaviour* **24**, 398–414.
4.3.2, 4.4

Halliday T.R. (1977) The effect of experimental manipulation of breathing behaviour on the sexual behaviour of the smooth newt, *Triturus vulgaris*. *Animal Behaviour* **25**, 39–45.
4.3.2

Halliday T.R. (1980) Motivational systems and interactions between activities. In: *The Analysis of Motivational Processes* (ed. F.M. Toates & T.R. Halliday), pp. 205–220. Academic Press, London.
5.8.2

Halliday T.R. & Sweatman H.P.A. (1976) To breathe or not to breathe; the newt's problem. *Animal Behaviour* **24**, 551–561.
4.4, 5.6.2

Hansen R.M. & Skavenski A.A. (1977) Accuracy of eye position information for motor control. *Vision Research* **17**, 919–926.
2.10

Harker J.E. (1974) The biological clock. *Science Progress* **61**, 175–189.
3.3.3

Harkness L.I.K. (1977) A behavioural study of chameleons. Unpublished D.Phil. thesis, Oxford University.
2.7

Harris L.R. (1980) The superior colliculus and movements of the head and eyes in cats. *Journal of Physiology* **300**, 367–391.
2.10.1

Hart M.'t (1978) A study of a short term behaviour cycle: creeping through in the three-spined stickleback. *Behaviour* **67**, 1–66.
4.4

Heiligenberg W. (1974) A stochastic analysis of fish behaviour. In: *Motivational Control Systems Analysis* (ed. D.J. McFarland), pp. 87–118. Academic Press, London.
5.3.2

Heiligenberg W. (1977) *Principles of Electrolocation and Jamming Avoidance in Electric Fish.* Springer-Verlag, Berlin.
1.1, 1.2.4

Heinrich B. (1979) *Bumblebee Economics.* Harvard University Press, Cambridge, Mass.
2.5

Hill D.E. (1979) Orientation by jumping spiders of the genus *Phiddipus* (Araneae: Salticidae) during the pursuit of prey. *Behavioral Ecology and Sociobiology* **5,** 301–322.
2.7

Hinde R.A. (1958) The nest-building behaviour of domesticated canaries. *Proceedings of the Zoological Society of London* **131,** 1–48.
4.4

Hinde R.A. (1970) *Animal Behaviour.* McGraw-Hill, New York.
3.2.1, 4.2.2, 4.5, 5.1, 5.3.2, 5.5.3, 5.6, 5.9

Hinde R.A. (1982) *Ethology.* Fontana, London.
4.2.2, 4.2.3, 4.3.2, 5.9

Hinde R.A. & Stevenson J.G. (1969) Sequences of behavior. *Advances in the Study of Behavior* **2,** 267–296.
4.4

Hinde R.A. & Stevenson J.G. (1970) Goals and response control. In: *Development and Evolution of Behavior* (ed. L.A. Aronson, E. Tobach, D.S. Lehrman & J.S. Rosenblatt), pp. 216–237. W.H. Freeman, San Francisco.
3.4.1

Hinde R.A. & Steel E. (1978) The influence of daylength and male vocalisations on the estrogen-dependent behavior of female canaries and budgerigars, with discussion of data from other species. *Advances in the Study of Behavior* **8,** 39–73.
4.4

Hogan J.A. (1980) Homeostasis and behaviour. In: *Analysis of Motivational Processes* (ed. F.M. Toates & T.R. Halliday), pp. 3–21. Academic Press, London.
4.6, 4.8

Hogan J.A., Kleist S. & Hutchings C.S.L. (1970) Display and food as reinforcers in the Siamese fighting fish (*Betta splendens*). *Journal of Comparative and Physiological Psychology* **70,** 351–357.
4.6.1

Holst E. von (1954) Relations between the central nervous system and the peripheral organs. *British Journal of Animal Behaviour* **2,** 89–94.
1.1

Holst E. von & St Paul U. von (1963) On the functional organisation of drives. *Animal Behaviour* **11,** 1–20.
5.5.3

Hopkins C.D. (1977) Electric communication. In: *How Animals Communicate* (ed. T.A. Sebeok), pp. 263–289. Indiana University Press, Bloomington.
1.2.4

Hörnicke H. & Batsch F. (1977) Caecotrophy in rabbits—a circadian function. *Journal of Mammalogy* **58**, 240.
    4.5.2
Houston A.I. (1980) Godzilla and the creature from the black lagoon. In: *The Analysis of Motivational Processes* (ed. F.M. Toates & T.R. Halliday), pp. 297–318. Academic Press, London.
    5.8.1, 5.9
Houston A.I. (1982) Transitions and time sharing. *Animal Behaviour* **30**, 615–625.
    5.6, 5.6.1, 5.9
Houston A.I., Halliday T.R. & McFarland D.J. (1977) Towards a model of the courtship of the smooth newt, *Triturus vulgaris*, with special emphasis on problems of observability in the simulation of behaviour. *Medical and Biological Engineering* **15**, 49–61.
    4.3.2
Howard I.P. (1982) *Human Visual Orientation*. John Wiley & Sons, Chichester.
    2.12
Hoyle G. (1976) Approaches to understanding the neurophysiological basis of behavior. In: *Simpler Networks and Behavior* (ed. J.C. Fentress), pp. 21–38. Sinauer, Sunderland, Mass.
    3.2.1
Hoyle G. (1978) Where did the notion of 'command neurons' come from? *The Behavioral and Brain Sciences* **1**, 10–11.
    3.4.2
Hughes B.O. (1972) A circadian rhythm of calcium intake in the domestic fowl. *British Poultry Science* **13**, 485–493.
    4.5.2
Hull C.L. (1943) *Principles of Behavior*. Appleton-Century-Crofts, New York.
Humphrey N.K. (1974) Vision in a monkey without striate cortex: a case study. *Perception* **3**, 241–255.
    2.8
Iersel J.J.A. van & Bol A.C.A. (1958) Preening of two tern species. A study of displacement activities. *Behaviour* **13**, 1–88.
    5.6, 5.6.1, 5.6.3
Ingle D. (1976) Spatial vision in anurans. In: *The Amphibian Visual System* (ed. K.V. Fite), pp. 119–141. Academic Press, New York.
    2.11
Isaac D. & Marler P. (1963) Ordering of sequences of singing behaviour of mistle thrushes in relationship to timing. *Animal Behaviour* **11**, 179–188.
    3.3.3
Johansson G. (1975) Visual motion perception. *Scientific American* **232** (6), 76–88.
    2.5
Kalmijn A.J. (1971) The electric sense of sharks and rays. *Journal of Experimental Biology* **55**, 371–383.
    1.2.4
Kater S.B. & Rowell C.H.F. (1973) Integration of sensory and centrally programmed components in generation of cyclical feeding activity of *Helisoma trivolvis*. *Journal of Neurophysiology* **36**, 142–155.
    3.1, 3.2.1
Katz P.L. (1974) A long term approach to foraging optimisation. *American Naturalist* **108**, 758–782.
    5.8.1

Kear J. (1967) Experiments with young nidifugous birds on a visual cliff. *Wildfowl Trust 18th Annual Report*, 122–124.
2.7

Kennedy D. & Davis W.J. (1977) The organisation of invertebrate motor systems. In: *Handbook of Physiology. Vol. 2. Neurophysiology* (ed. E.R. Kandel), pp. 1023–1087. American Physiological Society, Bethesda, Maryland.
3.4.2

Keverne E.B. (1976) Sexual receptivity and attractiveness in the female rhesus monkey. *Advances in the Study of Behavior* **7**, 155–200.
4.3.3

Kissilef H.R. (1969) Food associated drinking in the rat. *Journal of Comparative and Physiological Psychology* **67**, 284–300.
5.5.1

Knudsen E.I. (1981) The hearing of the barn owl. *Scientific American* **245** (6), 82–91.
2.8

Knudsen E.I. & Konishi M. (1978) Space and frequency are represented separately in auditory midbrain of the owl. *Journal of Neurophysiology* **41**, 870–884.
2.8

Koshland D.E. (1979) A model regulatory system: bacterial chemotaxis. *Physiological Reviews* **59**, 811–862.
2.2, 2.11

Koshland D.E. (1980) *Bacterial Chemotaxis as a Model Behavioral System*. Raven Press, New York.
2.12

Kovac M.P. & Davis W.J. (1977) Behavioural choice: neural mechanisms in *Pleurobranchaea. Science* **198**, 632–634.
5.2

Kovac M.P. & Davis W.J. (1980a) Reciprocal inhibition between feeding and withdrawal behaviours in *Pleurobranchaea. Journal of Comparative Physiology* **139**, 77–86.
2.11, 5.2

Kovac M.P. & Davis W.J.'(1980b) Neural mechanisms underlying behavioral choice in *Pleurobranchaea. Journal of Neurophysiology* **43**, 469–487.
2.11

Krebs J.R. & McCleery R.H. (1983) Foraging and time budgets. In: *Economics of Animal Behaviour* (ed. N.B. Davies & J.R. Krebs). Blackwell Scientific Publications, Oxford.
5.8.2

Krebs J.R., Stephens D.W. & Sutherland W.J. (1983) Perspectives in optimal foraging. In: *Perspectives in Ornithology* (ed. A.H. Brush & G.A. Clark). Cambridge University Press, New York.
5.8.2

Lack D. (1953) *The Life of the Robin*. Penguin Books, London.
1.1

Land M.F. (1972) Mechanisms of orientation and pattern recognition by jumping spiders. In: *Information Processing in the Visual Systems of Arthropods* (ed. R. Wehner), pp. 231–247. Springer-Verlag, Berlin & New York.
1.2.1

Land M.F. (1980a) Optics and vision in invertebrates. In: *Handbook of Sensory Physiology* Vol. VII/6B (ed. H. Autrum), pp. 471–592. Springer-Verlag, Berlin.
1.2.1

Land M.F. (1980b) Compound eyes: old and new optical mechanisms. *Nature* (London) **287**, 681–686.
1.2.1

Landau B., Gleitman H. & Spelke E. (1981) Spatial knowledge and geometric representation in a child blind from birth. *Science* **213**, 1275 -1278.
2.6

Larkin S.P.C. & McFarland D.J. (1978) The cost of changing from one activity to another. *Animal Behaviour* **26**, 1237–1246.
5.6.2

Lee D.N. (1974) Visual information during locomotion. In: *Perception: Essays in Honor of J.J. Gibson* (ed. R.B. McLeod & H.L. Pick Jr), pp. 250–267. Cornell University Press, Ithaca, New York.
2.5

Lee D.N. (1980) The optic flow field: the foundation of vision. *Philosophical Transactions of the Royal Society of London* B **290**, 169–179.
1.2.1

Lefebvre L. (1981) Grooming in crickets: timing and hierarchical organisation. *Animal Behaviour* **29**, 973–984.
3.4.2, 5.6.2

Lehrman D.S. (1961) The presence of the mate and of nesting material as stimuli for the development of incubation behaviour and for gonadotropin secretion in the ring dove (*Streptopelia risoria*). *Endocrinology* **68**, 507–516.
4.4

Le Magnen J. (1981) The metabolic basis of dual periodicity of feeding in rats. *Behavioral and Brain Sciences* **4**, 561–607.
6.3.4

Lindauer M. (1969) Behavior of bees under optical learning conditions. In: *Processing of Optical Data by Organisms and by Machines* (ed. W. Reichardt), pp. 527–543. Academic Press, New York.
1.1

Lissmann H.W. (1951) Continuous electrical signals from the tail of a fish, *Gymnarchus niloticus* Cuv. *Nature* (London) **167**, 201.
1.2.4

Lissmann H.W. (1963) Electric location by fishes. *Scientific American* **208** (3), 50–59.
1.1

Lock A. & Collett T. (1979) A toad's devious approach to its prey: a study of some complex uses of depth vision. *Journal of Comparative Physiology* **131**, 179–189.
2.6

Lock A. & Collett T. (1980) The three dimensional world of a toad. *Proceedings of the Royal Society of London* B **206**, 481–487.
2.6

Lorenz K. (1932) Betrachtungen über das Erkennen der arteigenen Triebhandlungen der Vögel. *Journal für Ornithologie* **80**, 50–98. (Reprinted in English in Lorenz 1970.)
3.2.1

Lorenz K. (1937) Über die Bildung des Instinktbegriffes. *Naturwissenschaften* **25**, 289–300, 307–318, 324–331. (Reprinted in English in Lorenz 1970.)
3.2.1

Lorenz K. (1950) The comparative method in studying innate behaviour patterns. *Symposia of the Society for Experimental Biology* **4**, 221–268.
Introduction, 4.2.4

Lorenz K. (1966) *On Aggression*. Methuen, London.
Introduction, 4.6.1

Lorenz K. (1970) *Studies in Animal and Human Behaviour*. Methuen, London.
3.2.1

Ludlow A.R. (1976) The behaviour of a model animal. *Behaviour* **58**, 131–172.
5.6.3

Ludlow A.R. (1980) The evolution and simulation of a decision maker. In: *The Analysis of Motivational Processes* (ed. F.M. Toates & T.R. Halliday), pp. 273–296. Academic Press, London.
5.6.3, 5.9, 6.7

McCleery R.H. (1978) Optimal behaviour sequences and decision making. In: *Behavioural Ecology* (ed. J.R. Krebs & N.B. Davies), pp. 377–410. Blackwell Scientific Publications, Oxford.
5.8.1, 5.9

McCosker J.E. (1977) Flashlight fishes. *Scientific American* **236** (3), 106–112.
1.2.4

McFarland D.J. (1966) On the causal and functional significance of displacement activities. *Zeitschrift für Tierpsychologie* **23**, 217–235.
5.4, 5.6.3

McFarland D.J. (1969) Mechanisms of behavioural disinhibition. *Animal Behaviour* **17**, 238–242.
5.6

McFarland D.J. (1971) *Feedback Mechanisms in Animal Behaviour*. Academic Press, London.
6.9

McFarland D.J. (1974) Time sharing as a behavioral phenomenon. In: *Advances in the Study of Behavior* **5**, 201–225.
5.6.2

McFarland D.J. (1976) Form and function in the temporal organisation of behaviour. In: *Growing Points in Ethology* (ed. P.P.G. Bateson & R.A. Hinde), pp. 59–93. Cambridge University Press, Cambridge.
4.3.1

McFarland D.J. & McFarland F.J. (1968) Dynamic analysis of an avian drinking response. *Medical and Biological Engineering* **6**, 659–668.
5.6.2

McFarland D.J. & Rolls B.J. (1972) Suppression of feeding by intracranial injections of angiotensin. *Nature* (London) **236**, 172–173.
5.5.2

McFarland D.J. & Sibly R.M. (1975) The behavioural final common path. *Philosophical Transactions of the Royal Society of London* B **270**, 265–293.
5.1, 5.3, 5.7, 6.3.3

McFarland D.J. & Houston A.I. (1981) *Quantitative Ethology: The State Space Approach*. Pitman, London.
4.3.1, 5.5, 5.7, 5.8.1, 5.9, 6.9

McNab R.M. & Ornston M.K. (1977) Normal to curly flagellar transitions and their role in bacterial tumbling. *Journal of Molecular Biology* **112**, 1–30.
2.2

Macevicz S. & Oster G. (1976) Modelling social insect populations II: Optimal reproductive strategies in annual eusocial insect colonies. *Behavioral Ecology and Sociobiology* **1**, 265–282.
5.8.1

Marler P. (1961) The filtering of external stimuli during instinctive behaviour. In: *Current Problems in Animal Behaviour* (ed. W.H. Thorpe & O.L. Zangwill), pp. 150–166. Cambridge University Press, Cambridge.
1.3

Marr D.C. (1982) *Vision.* W.H. Freeman, San Francisco.
Introduction

Maynard Smith J. (1978) Optimisation theory in evolution. *Annual Review of Ecology and Systematics* **9**, 31–56.
5.8.2

Meddis R. (1975) On the function of sleep. *Animal Behaviour* **23**, 676–691.
4.5.2

Menzel R. (1979) Spectral sensitivity and color vision in invertebrates. In: *Handbook of Sensory Physiology* Vol. VII/6A (ed. H. Autrum), pp. 503–580. Springer-Verlag, Berlin.
1.2.1

Merzenich M.M., Roth G.L., Andersen R.A., Knight P.L. & Colwell S.A. (1977) Some basic features of organization of the central auditory nervous system. In: *Psychophysics and Physiology of Hearing* (ed. E.F. Evans & J.P. Wilson), pp. 485–497. Academic Press, London.
2.8

Messenger J.B. (1981) Comparative physiology of vision in molluscs. In: *Handbook of Sensory Physiology.* Vol. VII/6C (ed. H. Autrum), pp. 93–200. Springer-Verlag, Berlin.
1.2.1

Metz H. (1974) Stochastic models for the temporal fine structure of behaviour sequences. In: *Motivational Control Systems Analysis* (ed. D.J. McFarland), pp. 5–86. Academic Press, London.
3.3.2

Meyer-Rochow V.B. (1974) Structure and function of the larval eye of the saw fly *Perga* (Hymenoptera). *Journal of Insect Physiology* **20**, 1565–1591.
1.2.1

Michael R.P. & Bonsall R.W. (1977) Periovulatory synchronisation of behaviour in male and female rhesus monkeys. *Nature* (London) **265**, 463–464.
4.3.3

Michael R.P. & Zumpe D. (1981) Environmental influences on the sexual behaviour of rhesus monkeys. In: *Environmental Factors in Mammalian Reproduction* (ed. D. Gilmore & B. Cook), pp. 77–99. Macmillan, London.
4.3.3

Michelsen A. (1978) Sound reception in different environments. In: *Sensory Ecology* (ed. M.A. Ali), pp. 345–373. Plenum Press, New York.
1.2.3

Miles F.A. & Lisberger S.G. (1981) Plasticity in the vestibulo-ocular reflex: a new hypothesis. *Annual Review of Neuroscience* **4**, 273–299.
2.11

Miller G.A., Galanter E. & Pribram K.H. (1960) *Plans and the Structure of Behavior.* Holt, Rinehart & Winston, New York.
3.4.2

Miller N.E. (1956) Effects of drugs on motivation: the value of using a variety of measures. *Annals of the New York Academy of Sciences* **65**, 318–333.
4.2.2

Miller N.E. (1959) Liberalisation of basic S-R concepts: extensions to conflict, behavior, motivation and social learning. In: *Psychology, A Study of a Science,* Study 1, Vol. 2 (ed. S. Koch), pp. 196–292. McGraw-Hill, New York.
4.2.2

Moltz H. (1965) Contemporary instinct theory and the fixed action pattern. *Psychological Reviews* **72,** 27–47.
3.2.1

Morasso P., Bizzi E. & Dichgans J. (1973) Adjustment of saccade characteristics during head movements. *Experimental Brain Research* **16,** 492–500.
2.9

Morrison S.D. (1968) Regulation of water intake of rats deprived of food. *Physiology and Behavior* **3,** 75–81.
5.4, 5.5.1

Nelson K. (1964) The temporal patterning of courtship behaviour in the Glandulocaudine fishes (Ostariophysi, Characidae). *Behaviour* **24,** 90–146.
3.3.3

Nelson K. (1973) Does the holistic study of behavior have a future? In: *Perspectives in Ethology* (ed. P.P.G. Bateson & P.H. Klopfer), pp. 281–328. Plenum Press, New York.
3.4.2

Neuweiler G. & Möhres F.P. (1967) Die Rolle des Ortungsgedächtnisses bei der Oreintierung des Grossblatt-Fledermaus *Megaderma lyra. Zeitschrift für vergleichende Physiologie* **57,** 147–171.
2.6

Newman E.A. & Hartline P.H. (1982) The infrared 'vision' of snakes. *Scientific American* **246** (3), 98–107.
2.8

Oatley K. (1978) *Perceptions and Representations.* Methuen, London.
3.5, 5.2

Oatley K. & Toates F.M. (1969) The passage of food through the gut of rats and its uptake of food. *Psychonomic Science* **16,** 225–226.
5.5.1, 5.5.2

Oatley K. & Tonge D.A. (1969) The effects of hunger on water intake in rats. *Quarterly Journal of Experimental Psychology* **21,** 162–171.
5.4

Oatley K. & Toates F.M. (1973) Osmotic inhibition of eating as a subtractive process. *Journal of Comparative and Physiological Psychology* **82,** 268–277.
5.5.2

O'Keefe J. & Nadel L. (1978) *The Hippocampus as a Cognitive Map.* Clarendon Press, Oxford.
6.6

O'Kelly L.I., Falk J.L. & Flint D. (1958) Water regulation in the rat: I. Gastrointestinal exchange rates of water and sodium chloride in thirsty animals. *Journal of Comparative and Physiological Psychology* **51,** 16–21.
6.4

Ollason J.C. & Slater P.J.B. (1973) Changes in the behaviour of the male zebra finch during a twelve-hour day. *Animal Behaviour* **21,** 191–196.
4.5

Olton D.S. & Samuelson R.J. (1976) Remembrance of places passed: spatial memory in rats. *Journal of Experimental Psychology: Animal Behavior Processes* **2**, 97–116.
3.4.1

Oster G.F. & Wilson E.O. (1978) *Caste and Ecology in the Social Insects*. Princeton University Press, Princeton, New Jersey.
5.8.1

Palmer J.D. (1975) Biological clocks of the tidal zone. *Scientific American* **232** (2), 70–79.
3.3.3

Panksepp J., Toates F.M. & Oatley K. (1972) Extinction induced drinking in hungry rats. *Animal Behaviour* **20**, 493–498.
5.4, 5.6.3

Peters P.J. (1970) Orb web construction: interaction of spider (*Araneus diadematus* CL.) and thread configuration. *Animal Behaviour* **18**, 478–484.
3.4.1

Peters R S (1960) *The Concept of Motivation*. Routledge & Kegan Paul, London.
5.1

Pirenne M.H. (1967) *Vision and the Eye*. Chapman & Hall, London.
1.2.1

Polit A. & Bizzi E. (1979) Characteristics of the motor programs underlying arm movements in monkeys. *Journal of Neurophysiology* **42**, 183–194.
2.10.2

Pöppel E., Held R. & Frost D. (1973) Residual visual function after brain wounds involving the central visual pathways in man. *Nature* (London) **243**, 295–296.
2.8

Porten K. van der, Redmann G., Rothman B. & Pinsker H. (1980) Neuroethological studies of freely swimming *Aplysia brasiliana*. *Journal of Experimental Biology* **84**, 245–257.
3.3.3

Powers W.T. (1978) Quantitative analysis of purposive systems: some spadework at the foundations of scientific psychology. *Psychological Review* **85**, 417–435.
6.6

Prosser C.L. (1973) *Comparative Animal Physiology* W.B. Saunders, Philadelphia
1.4

Purcell E.M. (1977) Life at low Reynolds number. *American Journal of Physics* **45**, 3–11.
2.2

Rashevsky N. (1960) *Mathematical Biophysics, Vol. 2*, 3rd edn. Dover Publications, New York.
5.8

Reichardt W. & Poggio T. (1976) Visual control of orientation behaviour in the fly. Part 1. A quantitative analysis. *Quarterly Review of Biophysics* **9**, 311–375.
2.5

Reingold S.C. & Gelperin A. (1980) Feeding motor programme in *Limax*. II. Modulation of sensory inputs in intact animals and isolated central nervous systems. *Journal of Experimental Biology* **85**, 1–19.
3.2.1

214 *References*

Rijnsdorp A., Daan S. & Dijkstra C. (1981) Hunting in the kestrel, *Falco tinnunculus,* and the adaptive significance of daily habits. *Oecologia* **50,** 391–406.
4.5.2

Robinson D.A. (1970) Oculomotor unit behavior in the monkey. *Journal of Neurophysiology* **33,** 393–404.
2.10.2

Robinson D.A. (1972) Eye movements evoked by collicular stimulation. *Vision Research* **12,** 1795–1808.
2.10.1

Robinson D.A. (1981) The use of control systems analysis in the neurophysiology of eye movements. *Annual Review of Neuroscience* **4,** 463–503.
2.10

Rodgers R.S. & Roseburgh R.D. (1979) Computing a grammar for sequences of behavioural acts. *Animal Behaviour* **27,** 737–749.
3.4.2

Roeder K.D. (1965) Moths and ultrasound. *Scientific American* **212** (4), 94–104.
1.2.3

Rolls B.J. & McFarland D.J. (1973) Hydration releases inhibition of feeding produced by intracranial angiotensin. *Physiology and Behavior* **11,** 881–884.
5.5.2

Rolls B.J. & Rolls E.T. (1982) *Thirst.* Cambridge University Press, Cambridge.
5.5.2

Rolls B.J. & Wood R.J. (1977) The role of angiotensin in thirst. *Pharmacology Biochemistry and Behavior* **6,** 245–250.
5.5.2

Rolls B.J., Wood R.J. & Rolls E.T. (1980) Thirst: the initiation, maintenance and termination of drinking. In: *Progress in Psychobiology and Physiological Psychology, Vol. 9* (ed. J.M. Sprague & A.N. Epstein), pp. 263–321. Academic Press, New York.
5.5.2

Roper T.J. (1973) Nesting material as a reinforcer for female mice. *Animal Behaviour* **21,** 733–740.
4.6.1

Roper T.J. (1975) Nest material and food as reinforcers for fixed-ratio responding in mice. *Learning and Motivation* **6,** 327–343.
4.6.1

Roper T.J. (1980) 'Induced' behaviour as evidence for non-specific motivational effects. In: *The Analysis of Motivational Processes* (ed. F.M. Toates & T.R. Halliday), pp. 221–242. Academic Press, London.
4.5, 5.4, 5.9

Roper T.J. (1981) What is meant by 'schedule induced' and how general is schedule induction. *Animal Learning and Behavior* **9,** 433–440.
5.4

Roper T.J. (1983) Learning as an adaptive phenomenon. In *Animal Behaviour, Vol. 3: Genes, Development and Learning* (ed. T.R. Halliday & P.J.B. Slater), pp. 178–212. Blackwell Scientific Publications, Oxford.
6.3.4

Roper T.J. & Crossland G. (1982) Mechanisms underlying feeding and drinking transitions in rats. *Animal Behaviour* **30,** 602–614.
5.6

Rossel S. (1980) Foveal fixation and tracking in the praying mantis. *Journal of Comparative Physiology* **139**, 307–331.
2.5

Rovner J.S. (1967) Acoustic communication by a lycosid spider *Lycosa rabida*, Walckenaer. *Animal Behaviour* **15**, 273–281.
1.2.3

Rowell C.H.F. (1961) Displacement grooming in the chaffinch. *Animal Behaviour* **9**, 38–63.
5.6, 5.6.1, 5.6.2, 5.6.3

Sachs B.D. & Barfield R.J. (1976) Functional analysis of masculine copulatory behavior in the rat. *Advances in the Study of Behavior* **7**, 91–154.
6.5

Sales G. & Pye D. (1974) *Ultrasonic Communication by Animals*. Chapman & Hall, London.
1,1, 1.2.3

Saunders D.S. (1976) The biological clock of insects. *Scientific American* **234** (2), 114–121.
3.3.3

Saunders D.S. (1977) *An Introduction to Biological Rhythms*. Blackie, Glasgow.
3.3.3

Schiller P.H. & Stryker M. (1972) Single unit recording and stimulation in the superior colliculus of the alert monkey. *Journal of Neurophysiology* **35**, 915–924.
2.10.1

Schiller P.H., True S.D. & Conway J.L. (1980) Deficits in eye movements following frontal eye field and superior colliculus ablations. *Journal of Neurophysiology* **44**, 1175–1189.
2.10

Schleidt W.M. (1964) Uber das Wirkungsgefüge von Balzbewegungen des Truthahnes. *Naturwissenschaften* **51**, 445–446.
3.3.2

Schleidt W.M. (1974) How 'fixed' is the fixed action pattern? *Zeitschrift für Tierpsychologie* **36**, 184–211.
3.2.2

Schmidt R.F. (ed.) (1978) *Fundamentals of Sensory Physiology*. Springer-Verlag, Berlin & New York.
1.1.1, 1.4

Schmidt R.S. (1974) Neural correlates of frog calling. Independence from peripheral feedback. *Journal of Comparative Physiology* **88**, 321–333.
3.2.1

Schneider D. (1974) The sex-attractant receptor of moths. *Scientific American* **231** (1), 28–35.
1.2.5

Schnitzler H.U. & Henson O.W. (1980) Performance of airborne animal sonar systems. 1. Microchiroptera. In: *Animal Sonar Systems* (ed. R.-G. Busnel & J.F. Fish), pp. 109–181. Plenum Press, New York & London.
2.6

Sevenster P. (1961) A causal analysis of a displacement activity (fanning in *Gasterosteus aculeatus* L.). *Behaviour* Suppl. **9**, 1–170.
5.6, 5.6.1, 5.6.2

Sevenster-Bol A.C.A. (1962) On the causation of the drive reduction of a consummatory act. *Archives néerlandaises de Zoologie* **15**, 175–236.
4.4

Sherry D.F., Mrosovsky N. & Hogan J.A. (1980) Weight loss and anorexia during incubation in birds. *Journal of Comparative and Physiological Psychology* **94**, 89–98.
4.6.1

Sherry D.F., Krebs J.R. & Cowie R.J. (1981) Memory for the location of stored food in marsh tits. *Animal Behaviour* **29**, 1260–1266.
3.4.1

Shorey H.H. (1977) Pheromones. In: *How Animals Communicate* (ed. T.A. Sebeok), pp. 137–163. Indiana University Press, Bloomington.
1.2.5

Sibly R.M. & McFarland D.J. (1974) A state space approach to motivation. In: *Motivational Control Systems Analysis* (ed. D.J. McFarland), pp. 213–250. Academic Press, New York.
5.7

Sibly R.M. & McFarland D.J. (1976) On the fitness of behavior sequences. *American Naturalist* **110**, 610–617.
5.8.1

Silver R. (1978) The parental behavior of ring doves. *American Scientist* **66**, 209–215.
4.5.1

Simpson S.J. (1981) An oscillation underlying feeding and a number of other behaviours in fifth-instar *Locusta migratoria* nymphs. *Physiological Entomology* **6**, 315–324.
4.5.2

Simpson S.J. (1982) Patterns in feeding: a behavioural analysis using *Locusta migratoria* nymphs. *Physiological Entomology* **7**, 325–336.
4.5.2

Slater P.J.B. (1973) Describing sequences of behavior. In: *Perspectives in Ethology*, Vol. 1 (ed. P.H. Klopfer & P.P.G. Bateson), pp. 131–153. Plenum Press, New York.
3.3.2, 4.4

Slater P.J.B. (1978a) *Sex Hormones and Behaviour*. Edward Arnold, London.
4.5.1

Slater P.J.B. (1978b) Data collection. In: *Quantitative Ethology* (ed. P.W. Colgan), pp. 7–24. John Wiley & Sons, New York.
5.1

Slater P.J.B. & Ollason J.C. (1972) The temporal pattern of behaviour in isolated male zebra finches: transition analysis. *Behaviour* **42**, 248–269.
3.3.2

Slater P.J.B. & Wood A.M. (1977) Does activation influence short-term changes in zebra finch behaviour? *Animal Behaviour* **25**, 736–746.
4.5

Sparks D.L., Holland R. & Guthrie B.L. (1976) Size and distribution of movement fields in the monkey superior colliculus. *Brain Research* **113**, 21–34.
2.10.1

Srinivasan M.V. (1977) A visually evoked roll response in the house-fly. Open-loop and closed-loop studies. *Journal of Comparative Physiology* **119**, 1–14.
2.5

Stamps J.A. & Barlow G.W. (1973) Variation and stereotypy in the displays of *Anolis aeneus* (Sauria: Iguanidae). *Behaviour* **47**, 67–94.
    3.2.1
Stange G. & Howard J. (1979) An ocellar dorsal light response in a dragonfly. *Journal of Experimental Biology* **83**, 351–355.
    2.5
Stein B.E. & Clamann H.P. (1981) Control of pinna movements and sensorimotor register in cat superior colliculus. *Brain, Behavior and Evolution* **19**, 180–192.
    2.10.1
Stein P.S.G. (1978) Motor systems, with specific reference to the control of locomotion. *Annual Review of Neuroscience* **1**, 61–81.
    3.2.1
Stricker E.M. (1973) Thirst, sodium appetite, and complementary physiological contributions to the regulation of intravascular fluid volume. In *The Neuropsychology of Thirst: New Findings and Advances in Concepts* (ed. A.N. Epstein, H.R. Kissileff & E. Stellar), pp. 73–98. V.H. Winston, Washington.
    6.4
Suga N., Kuzirai K. & O'Neill W.E. (1981) How biosonar information is represented in the bat cerebral cortex. In: *Neuronal Mechanisms of Hearing* (ed. J. Syka & L. Aitkin), pp. 197–219. Plenum Press, New York.
    2.8
Thompson T. & Bloom W. (1966) Aggressive behavior and extinction induced response increase. *Psychonomic Science* **5**, 335–336.
    5.4
Thomson J.A. (1980) How do we use information to control locomotion? *Trends in Neurosciences* **3**, 247–250.
    2.6
Thorpe W.H. (1961) *Bird Song*. Cambridge University Press, Cambridge.
    1.1
Thorpe W.H. (1963) *Learning and Instinct in Animals*. Methuen, London.
    3.4.1
Tinbergen N. (1951) *The Study of Instinct*. Clarendon Press, Oxford.
    Introduction, 3.4.2, 4.4
Tinbergen N. (1959) Comparative studies of the behaviour of gulls (Laridae): a progress report. *Behaviour* **15**, 1–70.
    1.2.1
Toates F.M. (1975) *Control Theory in Biology and Experimental Psychology*. Hutchinson Educational, London.
    6.2.1, 6.9
Toates F.M. (1979a) Homeostasis and drinking. *Behavioral and Brain Sciences* **2**, 95–139.
    5.5.1
Toates F.M. (1979b) Water and energy in the interaction of thirst and hunger. In: *Chemical Influences on Behaviour* (ed. K. Brown & S.J. Cooper), pp. 135–200. Academic Press, London.
    4.5.2, 6.8
Toates F.M. (1980a) *Animal Behaviour. A Systems Approach*. John Wiley & Sons, Chichester.
    4.6, 4.6.1, 4.8, 5.5.1, 5.9, 6.3.4, 6.4, 6.9

Toates F.M. (1980b) A systems approach to sexual behaviour. In: *Analysis of Motivational Processes* (ed. F.M. Toates & T.R. Halliday), pp. 319–338. Academic Press, London.
6.5

Toates F.M. (1981) The control of ingestive behaviour by internal and external stimuli—a theoretical review. *Appetite* **2**, 35–50.
5.8.2, 6.4, 6.7

Toates F.M. & Oatley K. (1970) Computer simulation of thirst and water balance. *Medical and Biological Engineering* **8**, 71–87.
6.4

Toates F.M. & Oatley K. (1972) Inhibition of *ad libitum* feeding in rats by salt injections and water deprivation. *Quarterly Journal of Experimental Psychology* **24**, 215–224.
5.5.1, 5.5.2

Toates F.M. & O'Rourke C. (1978) Computer simulation of male rat sexual behaviour. *Medical and Biological Engineering and Computing* **16**, 98–104.
6.5

Toates F.M. & Halliday T.R. (eds) (1980) *Analysis of Motivational Processes*. Academic Press, London.
4.8, 5.9, 6.9

Todd J.H. (1971) The chemical language of fishes. *Scientific American* **224** (5), 99–108.
1.2.5

Truman J.W. (1978) Hormonal release of stereotyped motor programmes from the isolated nervous system of the *Cercropia* silkmoth. *Journal of Experimental Biology* **74**, 151–173.
3.2.1

Truman J.W. & Sokelove P.G. (1972) Silk moth eclosion: Hormonal triggering of a centrally programmed pattern of behavior. *Science* **175**, 1491–1493.
3.2.1

Uexküll J. von (1957) A stroll through the worlds of animals and men. Reprinted and translated in: *Instinctive Behaviour, the Development of a Modern Concept* (ed. C.H. Schiller), pp. 5–80. Methuen, London.
1.1, 2.6

Viviani P. & Terzuolo C. (1980) Space–time invariance in learned motor skills. In: *Tutorials in Motor Behaviour* (ed. G.E. Sternbach & J. Requin), pp. 525–533. North-Holland, Amsterdam.
2.9

Vowles D.M. (1970) Neuroethology, evolution and grammar. In: *Development and Evolution of Behavior* (ed. L.R. Aronson, E. Tobach, D.S. Lehrman & J.S. Rosenblatt), pp. 194–215. W.H. Freeman, San Francisco.
3.4.2

Wagner A.R. (1959) The role of reinforcement and non-reinforcement in an 'apparent frustration effect'. *Journal of Experimental Psychology* **57**, 130–136.
5.4

Wagner H. (1982) Flow field variables trigger landing in flies. *Nature* (London) **297**, 147–148.
2.5

Wald G., Brown P.K. & Gibbons I.R. (1962) Visual excitation: a chemo-anatomical study. *Symposia of the Society for Experimental Biology* **16**, 32–57.
1.2.1

References 219

Walk R.D. (1978) Depth perception and experience. In: *Perception and Experience* (ed. R.D. Walk & H.L. Pick Jr.), pp. 77–103. Plenum Press, New York.
2.7

Wallace G.K. (1959) Visual scanning in the desert locust *Schistocerca gregaria*. *Journal of Experimental Biology* **36**, 512–525.
2.5

Wehner R. (1976) Polarised-light navigation by insects. *Scientific American* **235** (1), 106–115.
1.2.1

Wehner R. (1981) Spatial vision in arthropods. In: *Handbook of Sensory Physiology* Vol. VII/6C (ed. H. Autrum), pp. 287–616. Springer-Verlag, Berlin.
1.2.1, 2.5, 2.12

Whittington D.A., Hepp-Reymond M.C. & Flood W. (1981) Eye and head movements to auditory targets. *Experimental Brain Research* **41**, 358–363.
2.9

Wiepkema P.R. (1961) An ethological analysis of the reproductive behaviour of the bitterling. *Archives néerlandaises de Zoologie* **14**, 103–199.
5.3.2

Wiepkema P.R. (1968) Positive feedbacks at work during feeding. *Behaviour* **39**, 266–273.
5.6.2

Wiersma C.A.G. & Ikeda K. (1964) Interneurons commanding swimmeret movements in the crayfish, *Procambarus clarkii* (Girard). *Comparative Biochemistry and Physiology* **12**, 509–525.
3.4.2

Wiley R.H. (1973) The strut display of male sage grouse: a 'fixed' action pattern. *Behaviour* **47**, 129–152.
3.2.2

Willows A.O.D. (1971) Giant brain cells in mollusks. *Scientific American* **224** (2), 68–75.
3.2.1

Wilson E.O. (1963) Pheromones. *Scientific American* **208** (5), 100–114.
1.2.5

Wilson E.O. (1975) *Sociobiology: The New Synthesis*. Belknap Press, Cambridge, Mass.
Introduction

Wilson M. (1978) The functional organization of locust ocelli. *Journal of Comparative Physiology* **24**, 297–316.
2.5

Wilz K. (1970) Causal and functional analysis of dorsal pricking and nest activity in the courtship of the three spined stickleback *Gasterosteus aculeatus*. *Animal Behaviour* **18**, 682–687.
5.6.1, 5.6.2, 5.6.3

Winston P.H. & Brown R.H. (1979) *Artificial Intelligence*, Vol. 2. M.I.T. Press, Cambridge, Mass.
2.9

Wirtshafter D. & Davis J.D. (1977) Set points, settling points, and control of body weight. *Physiology and Behavior* **19**, 75–78.
6.3.2, 6.3.3

References

Wolf R. & Heisenberg M. (1980) On the fine structure of yaw torque in visual flight orientation of *Drosophila melanogaster* II. A temporally and spatially variable weighting function of the visual field ('visual attention'). *Journal of Comparative Physiology* **140**, 69–80.
    2.5
Wurtz R.H. & Albano J.E. (1980) Visual-motor function of the primate superior colliculus. *Annual Review of Neuroscience* **3**, 189–226.
    2.8, 2.10.1, 2.10.2
Yost W.A. & Nielsen D.W. (1977) *Fundamentals of Hearing*. Holt, Rinehart & Winston, New York.
    1.2.2

INDEX